INTERFACE: CALCULUS AND THE COMPUTER SECOND EDITION

David Smith

SAUNDERS COLLEGE PUBLISHING

Philadelphia   New York   Chicago
San Francisco   Montreal   Toronto
London   Sydney   Tokyo   Mexico City
Rio de Janeiro   Madrid

Address orders to:
383 Madison Avenue
New York, NY 10017

Address editorial correspondence to:
West Washington Square
Philadelphia, PA 19105

Text Typeface: 10/12 Times Roman
Compositor: Science Press
Acquisitions Editor: Leslie Hawke
Project Editor: Diane Ramanauskas
Copyeditor: Elaine L. Honig
Managing Editor & Art Director: Richard L. Moore
Art/Design Assistant: Virginia A. Bollard
Text Design: Emily Harste
Cover Design: Lawrence R. Didona
Production Manager: Tim Frelick
Assistant Production Manager: Maureen Iannuzzi

Cover credit: Photograph of French horn: Pete Turner/THE IMAGE BANK

Library of Congress Cataloging in Publication Data

Smith, David A.
    Interface : calculus and the computer.

    Bibliography: p.
    Includes index.

    1. Calculus—Data processing.   I. Title.
QA303.S596   1984   515'.028'54   83-20053
ISBN 0-03-070663-7

Interface: Calculus and the Computer   ISBN 0-03-070663-7

3456   032   987654321

CBS COLLEGE PUBLISHING
Saunders College Publishing
Holt, Rinehart and Winston
The Dryden Press

DEDICATION

The first edition of this book was dedicated
to my parents,
who equipped me for a journey
and let me choose where to go.

This edition is dedicated, with love,
to Dorothy,
my wife, my friend, my companion on the journey,
and a constant source of strength and inspiration.

PREFACE

The first edition of *Interface: Calculus and the Computer,* which appeared in 1976, was a little ahead of its time in that it predated widespread availability of microcomputers and time-sharing terminals for supplementing mathematics instruction with digital computation. It also predated the developing consensus among teachers of mathematics that the computer is here to stay, that its intelligent use requires an understanding of mathematics, and that the use of mathematics will increasingly take place in a computer environment. Now almost every college and university can provide computer services to its students, and many students and teachers have their own computers. Indeed, several colleges (more each year) are requiring or urging *every* student to own a computer and are developing creative ways to finance student purchases of microcomputers. It is safe to say that *now* is the time (for those who have not done so already) to enlarge on the computational component of the mathematics courses in the traditional curriculum. This book is a tool for doing that with the cornerstone of the curriculum—the standard calculus course. (That cornerstone role is currently under active debate. When a new "core" sequence is developed, with discrete and continuous mathematics in coequal roles, a number of "traditional" topics will disappear or be relegated to later courses for fewer students. Then the topics in this book may be presented a little differently, but they will still be there.)

Our access to computational tools has changed radically in the last eight years; however, our mathematics (at the lower division college level) has not changed significantly.

OBJECTIVES

This is a book about mathematics, not about computer science or programming, although the user will be expected to use a computer to do many of the exercises. Among "students of mathematics," I definitely include future users of mathematics, not just future mathematicians. Among "teachers," I include both those who have and those who have not previously used a computer in connection with a mathematics course. The *Instructor's Manual* accompanying this text provides help for those new to computing, plus other features (described later) that will be useful to all instructors.

This volume belongs to the genre of "supplementary computer-calculus texts," but I prefer to describe it as an "enrichment text" rather than as a "supplement." Its objectives are:

1. To provide a flexible supplement to any standard first-year calculus book to make possible a computer-based "laboratory" experience;

2. To introduce the student to applications of calculus to digital computation in a scientific setting;

3. To use the computer to illustrate the ideas normally encountered in the calculus course; and

4. To place the ideas discussed in appropriate historical context so that the student can see mathematics as a living and growing body of ideas.

COMPUTER FACILITIES REQUIRED

The best way to implement this text is with microcomputers, but a time-sharing minicomputer or mainframe with interactive terminals will do almost as well. Most of the exercises can also be carried out in a batch process environment, and a majority can be done (with greater effort) with a calculator. Samples of all the programs the students will be expected to write, plus several "canned" graphics programs, are provided on diskette for several popular microcomputers. The sample programs and "character graphics" versions of the canned programs are listed in the *Instructor's Manual* for users of other systems. The computer work may be done either in scheduled lab periods or *ad lib,* with students working alone or with two or three to a keyboard, supervised or unsupervised. (Novices will need some help at the start, but this may be provided just as well by more experienced students as by the instructor.)

PROGRAMMING LANGUAGE

For good or bad, BASIC is the *lingua franca* of instructional computing (just as FORTRAN still plays that role for scientific computing). Although computer scientists may disapprove of its lack of structure (and properly so, even though a structured version of BASIC is "just around the corner"), it is the only language that may be fairly described as universally available for instructional purposes. Furthermore, structure is not much of an issue with the 10- to 20-line programs that students will use to solve the exercises in this book. Therefore, the programming in this text will be in BASIC. Beyond the introductory chapters, the text is language-independent, with all algorithms expressed in a simple flow-charting scheme. Students who know or are learning another computer language are encouraged to use it instead.

This book has also been used in a "nonprogramming" environment (by several instructors at Duke University) with the aid of a "general problem solver" called MATHPROGRAM. With this tool, students need only learn how to formulate each mathematical algorithm as a short list of assignment and/or function definition statements, all of which are very close to the mathematical formulas themselves. MATHPROGRAM is available in BASIC for time-sharing computers and several popular microcomputers from CONDUIT, The University of Iowa, Iowa City, Iowa 52242.

COURSE STRUCTURES

I used various versions of this book for several years before its first publication, and many other instructors (and I) have used the first edition, primarily for a supplementary or laboratory course attached to or paralleling a standard course in single-variable calculus (through infinite series). If the standard course meets three hours per week for a full academic year, the computer component would normally meet an additional hour per week. However, the text can also be used with a variety of other classroom formats, such as:

1. One or two hours a week in a four- or five-hour calculus course for a full year;

2. A complete three-semester-hour course parallel to the second semester of calculus (perhaps as an introduction to numerical analysis, which might be preceded by a course in programming);

3. In a supplementary mode, but only for the second semester or second and third quarters of an academic year;

4. By careful topic selection, integrated into a three-hour calculus course; and

5. As a self-study text or projects manual for highly motivated or honors students.

Detailed suggestions for implementation in each of these modes may be found in the *Instructor's Manual*. The most difficult mode to implement is the three-hour course that cannot be expanded to four or five hours, therefore necessitating hard choices about what to leave out. Briefly, one might select very basic topics from this book—e.g., *one* method for finding roots, numerical integration only through the Trapezoidal Rule, perhaps one unit on polynomial approximation, the limit definition for e^x, and first-order, initial-value problems—presenting a new idea only every other week (on the average), instead of every week. Some room could be made in the syllabus by giving careful thought to the necessity for covering any or all of the following topics that may now be in your course: detailed development of the real number system, theory of continuous functions, more advanced methods of formal integration, and multiplicity of "applications" of the integral. If you cover either series or partial derivatives, could they wait until the second year? A few topics will fit without making much room. For example, root finding is very natural when discussing max/min problems; evaluation of upper and lower sums can (indeed, should) replace evaluation of $\int_0^b x\,dx$ and $\int_0^b x^2\,dx$ by summation formulas; if series are covered, the ratio-test evaluation method requires very little additional discussion. In a three-hour course, one might find it necessary to require a few additional lectures (perhaps evenings) on programming at the beginning of the course.

STRUCTURE OF THE BOOK

A glance at the Contents will show a "chapter" of this book is a very small unit. This arrangement provides maximum flexibility in matching the content and pace of the supplementary material to that of the basic course. A typical week's work might involve study of one chapter presenting a new idea, with one lecture on that topic, plus use of another chapter as "lab manual" for using the computer, which may be done on either a scheduled or unscheduled basis, supervised or unsupervised. In some weeks there will also be an optional follow-up chapter, to be read by serious students of mathematics after the laboratory experience.

Chapters 1 and 2 introduce the novice to programming in BASIC, and Chapters 4 and 5 provide the handful of programming techniques that will be needed to solve the exercises in the rest of the book. This is the material that may be read through quickly or skipped entirely by those who already know how to program or are using another language. Chapter 6 describes a simple function tabulator and gives the details of the canned graphics programs. This introductory unit (through Chap. 6) may be covered during the first part of the calculus course, even though there will be little point of contact except for the concepts of function and graph.

After the students are comfortable with writing ten-line programs and seeing them run, they are exposed to the mathematical foundation for the rest of the text, namely, the definitions and a few examples of sequences and limits (Chap. 7). It is *not* assumed that sequences will be taken up in the basic course at this point, and the treatment presented here is not intended to replace that which would normally come later in the course. The objective is for the student to become comfortable with the idea of computing terms of a sequence to get a very good approximation to some number known to be a limit of the sequence. (For those who want to take up sequences at this point, optional Chaps. 9 and 11 are provided in case your basic text does not support this point of view.)

From this point on, the various topics discussed are largely independent and may be covered in any order. These topics are computation of π (Chaps. 12–14); roots of equations (15–20); max/min problems (21–23); integration (24–31); differentiation (32–36); transcendental functions (37–47); initial-value problems (48–51); and evaluation of series (52–57). Chapter 58 contains suggestions for projects that follow up on most of the previous sections. Within each broad topic, there is additional flexibility in selection and arrangement of coverage, as indicated in the prerequisite chart.

The following code is used to identify special features of various portions of the text (chapters, sections, and sometimes smaller units):

[L] *Laboratory session.* The student should read through this material before approaching the computer and should have it in hand during the session.

[O] *Optional material.* The emphasis here is on the student's option, not the instructor's (although I invariably find some of this material creeping into my lectures). Students who have a serious interest in mathematics for its own sake will be more interested in the optional sections than others will be.

[R] *Reading material only.* This indicates sections that should be read by all students in the course, but that should *not* be taught or tested. My experience indicates that students generally appreciate a break from the rigors of reading just mathematics.

Bibliographic references are given at the end of each chapter, as appropriate, and sometimes elsewhere in the text. In some cases these are sources for the ideas covered; in others, they point out possible directions for the ambitious student who would like to pursue the ideas further. All references to the Bibliography are given by author's name, and, where necessary to distinguish among multiple listings, by date.

Displayed formulas, sections, examples, and theorems are each numbered consecutively within chapters. When a theorem, for example, is cited in another chapter, it is referred to by its chapter number and the order of its appearance within that chapter.

No answers to exercises are included in the text proper. As previously noted, the *Instructor's Manual* and supplementary diskette (for supported micros) contain sample programs for all exercises; each contains a file of numerical answers as well. This permits each instructor to decide how much of this information to provide to students and whether to use computer-based files or handouts.

APPLICATIONS

Applications of calculus are the principal reason for the prominent place the subject occupies in the undergraduate curriculum. Although I have not attempted a comprehensive survey of nonmathematical applications, I have tried to adhere to *usefulness* as a criterion for inclusion of topics. Where natural to do so, I have used nonmathematical settings for certain subjects. For example, Chapter 45 uses compound-interest calculations as a setting for the limit formula for e^x, and Chapters 48 and 50 use population dynamics to introduce first-order differential equations. Other applications to economics, physics, chemistry, biology, and medicine appear (especially in exercises) in Chapters 1, 19, 21, 22, 43, 46, 49, 51, and 58 (especially Sects. 5 and 10).

NUMERICAL ANALYSIS

For those who feel that the book contains too much numerical analysis, I must express my opinion that this is one of the most significant, useful, and accessible areas within mathematics for applications of calculus, especially because many nonmathematical applications depend on numerical solutions. For those who find too little numerical analysis (e.g., deliberate avoidance of questions of numerical significance wherever possible), I hasten to add that this is not a substitute for a numerical analysis course, only an introduction to (and perhaps an incentive for taking) one, and not everything of importance in such a course is an application of calculus.

PREREQUISITE CHART

Chapters 0 through 6 are on programming; Chapters 7, 8, and 10, on sequences. After covering that basic material, only the principal idea chapters are listed (laboratory sessions and optional chapters are omitted). Arrows show the extent to which nonoptional parts of the indicated chapters depend on earlier chapters. Additional prerequisites for optional sections and chapters are listed in the *Instructor's Manual*.

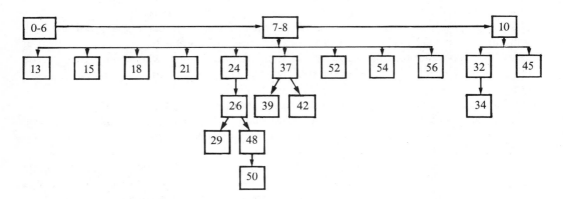

CALCULUS TEXTBOOK CROSS-REFERENCE CHARTS

Following the Bibliography are charts that match the chapters of this book, which are grouped by topic, to the appropriate sections of the following textbooks:

D. Berkey. *Calculus.* Saunders College Publishing, Philadelphia, 1984.

H. Anton. *Calculus with Analytic Geometry,* Second Edition. John Wiley & Sons, Inc., New York, 1984.

C. H. Edwards, Jr., and D. E. Penney. *Calculus and Analytic Geometry.* Prentice-Hall, Inc., Englewood Cliffs, NJ, 1982.

R. Ellis and D. Gulick. *Calculus with Analytic Geometry,* Second Edition. Harcourt Brace Jovanovich, New York, 1982.

P. Gillett. *Calculus and Analytic Geometry,* Second Edition. D. C. Heath & Co., 1984.

S. I. Grossman. *Calculus,* Second Edition., and *Calculus, Part 1.* Academic Press, Inc., New York, 1981.

R. E. Larson and R. P. Hostetler. *Calculus with Analytic Geometry,* Second Edition. D. C. Heath & Co., Lexington, MA, 1982.

L. Leithold. *The Calculus with Analytic Geometry,* Fourth Edition. Harper & Row Publishers, Inc., New York, 1981.

A. Mizrahi and M. Sullivan. *Calculus and Analytic Geometry.* Wadsworth Publishing, Belmont, CA, 1982.

S. L. Salas and E. Hille. *Calculus: One and Several Variables,* Fourth Edition. John Wiley & Sons, Inc., New York, 1982.

A. Shenk. *Calculus and Analytic Geometry,* Third Edition. Scott, Foresman & Co., Glenview, IL, 1984.

S. K. Stein. *Calculus and Analytic Geometry,* Third Edition. McGraw-Hill Book Company, New York, 1982.

E. W. Swokowski. *Calculus with Analytic Geometry,* Alternate Edition. Prindle, Weber & Schmidt, Inc., Boston, 1983.

E. W. Swokowski. *Calculus with Analytic Geometry,* Third Edition. Prindle, Weber & Schmidt, Inc., Boston, 1984.

G. B. Thomas, Jr., and R. L. Finney. *Calculus and Analytic Geometry,* Sixth Edition. Addison-Wesley Publishing Company, Reading, MA, 1984.

ACKNOWLEDGMENTS

This book would not have been possible without the assistance of many teachers over the last 30 years, and, it is a pleasure to acknowledge my indebtedness to them, even though I cannot name them all. These teachers include those whose courses and lectures I have attended; those who have been my colleagues; those known to me only through their published work; and especially those who have participated in my courses, and thereby taught me a great deal.

I am also indebted to Professors Leonard Gillman, Allen Ziebur, David Hayes, William Fuller, and Richard Williamson, who read portions of the manuscript for the first edition and offered constructive criticisms. Thanks are due to Paula Orzano, who assisted with the literature search, to Edward Cooley, who assisted with preparation of sample programs and solutions to exercises, and to the Duke University Undergraduate Research Assistantship Program, which paid for their services. Special thanks are due to Edith Minton, who typed the final manuscript under considerable pressure, and to the secretarial staff of the Duke University Mathematics Department, including Ann Davis, Polly Roberts, Ann Tunstall, Bonnie Ellis, and Judy Cohn, who assisted with preparation of several versions of the manuscript. The staff of the Duke University Computation Center has been extremely helpful, and I am particularly grateful to the late Harrison Register for stimulating my interest in conversational terminals as a teaching tool, which had much to do with the decision to offer the course from which this book has grown.

Many users of the first edition have been very helpful in providing suggestions for improvement, and I am especially grateful to Professors Frank Warner, Keith Stroyan, and Samuel Wagstaff for their comments. Through the marvel of home computers (mine happens to be a Commodore 64 with WordPro 3+ and a Star Gemini 10 printer), there is no typist to thank (or pay) for the second edition. To the nameless list of "teachers" previously mentioned, I must add dozens who have taught me about microcomputers over the last several years.

I would like to thank Leslie Hawke, Acquisitions Editor, Kate Pachuta, Editorial Assistant, and Diane Ramanauskas, Project Editor—all of Saunders College Publishing—for their assistance during the preparation and production of this text. I am also grateful to Emily Harste, who provided the design for this second edition.

Last, but surely not least, I am grateful for the support of my family, who have borne the burdens well known to families of authors, for the most part in good cheer.

D.A.S.

CONTENTS

CHAPTER 0

INTRODUCTION

0.1 CALCULUS AND THE COMPUTER

Since the earliest times of recorded history, one of the principal motivating forces for the development of mathematical concepts has been the need to *compute*—e.g., to compute the value of a commodity to be bought or sold or to compute the orbit of a planet. For several millennia mathematicians have labored diligently to devise systems of computation (as well as the supporting theories for them) to lessen the burden of carrying out computations that, when posed in the most obvious way, would lead to immense amounts of very tedious arithmetic, in addition to the even greater burden of simply *not knowing* how to compute a desired quantity.

The body of mathematics commonly known as *calculus* is such a computational system. Indeed, the word "calculus" *means* a system of computation and is derived from the Latin word of the same spelling meaning *stone*; one of the early calculating devices was a collection of stones used for counting. (Some related English words are *calculate, calcify, calcium,* and *calculus* as used in dentistry.) As commonly used in mathematics, calculus is an abbreviation for "differential and integral calculus," or, as Leibniz described the subject in the late seventeenth century, *calculus differentialis* and *calculus summatorius,* computations with differences and sums.[1] This subject has its origins in antiquity and has developed during the last 2000 years under the dual stimuli of finding out how to compute things—such as velocities, tangential directions, areas, and volumes—and how to simplify the system of computation. (A prime example of such simplification is now known as the Fundamental Theorem of Calculus, which is the connecting link between differential calculus and integral calculus.)

Most of what is studied now in a first course in calculus was known in the eighteenth century, and developments since then have been largely refinements, although important refinements, in shaping the subject as we know it today. However, during the past 40 years an important *nonmathematical* happening—namely, the rapid development of electronic computers—has greatly altered the course of mathematical development. In particular, it has become no longer necessary to fear computations requiring vast amounts of tedious arithmetic when machines are accessible that will do millions of arithmetic operations in a second and hardly ever make a mistake. This not only eliminates one of the two burdens mentioned previously, it fundamentally alters the other, in that it permits a broader range of possibilities for *knowing how* to solve a problem.

Because the introduction of fast computers represents a counterthrust to the motivation

---

[1]Gottfried Wilhelm Leibniz (1646–1716), German philosopher and mathematician, was one of the two people credited with the contemporaneous and independent development of calculus. The other was Sir Isaac Newton (1642–1727), an English physicist and mathematician. See Rosenthal for more details concerning the roles of these two men in the development of calculus. (References by author's name are to the Bibliography.)

behind calculus—i.e., the need to *avoid* computation—one might think it less important than it once was to study calculus. On the contrary, there is an important interaction between calculus and digital computation in which each can be applied to, and thereby enhance, the other. This interaction is the subject of this book. Put in simple terms, calculus tells us a great deal about which things are worth computing and how to do so meaningfully, and the computer makes possible concrete illustrations of abstract concepts of calculus, thereby leading to greater understanding. *"The purpose of computing is insight, not numbers."*[2]

In the early chapters of this book, we focus on learning to use the computer as a tool for getting certain jobs done. Presumably, you will study these chapters at the start of your calculus course, perhaps the first six weeks, and during this time you will probably not see any point of contact with calculus. Nevertheless, it is hoped that you will find the topics interesting in their own right, and you may rest assured that there will be plenty of interaction with calculus in subsequent chapters.

If you have not had any previous experience in working with a computer, you probably want to know what to expect. So at the conclusion of this brief introduction, we discuss computers in general and the one you will be using. Chapter 1 suggests some very simple computational problems to have in mind when you first try out the computer, and Chapter 2 serves as a "lab manual" for your first "hands-on" experience with the computer.

0.2 GENERALITIES ABOUT DIGITAL COMPUTERS

Viewed most superficially, a computer is a machine that does arithmetic very rapidly and very accurately—sort of a super version of a handheld or desk calculator. If we wanted to add a long list of numbers, say, 10,000 numbers, we could sit at the keyboard of a computer, enter the numbers one at a time, and accumulate their sum as they are entered. Of course, we could also do this with a calculator and in just about the same (long) time. A task of this sort does not take advantage of the computer's ability to add 10,000 numbers in a fraction of a second, because so much time is spent between additions communicating the terms of the sum to the machine. There are, however, two possible ways the computer may have direct access to our list of 10,000 numbers *without* our intervention. They may be "stored" in some sort of memory device whose contents can be fed to the arithmetic device much more rapidly than we can push buttons, or the numbers may themselves be answers from some other computations that the computer is capable of carrying out as it goes along. A calculator does not have either of these capacities, except in a very limited sense that need not concern us here.

Given access to the numbers, "data," there is another distinguishing characteristic of the computer that is even more important: It is not necessary to tell the machine 10,000 times to "add the next number," as one would have to do with the desk calculator (pushing the add button). Rather, the machine can be given the instruction, "add the next number" *once,* together with the instruction, "repeat the previous instruction 9999 times." The sequence of the two instructions, together with other appropriate instructions, such as, "tell me the answer," can be prepared in advance and entered into the machine prior to the start of computation. Once the instructions have been entered, a direct command that says, in effect,

---

[2]From *Numerical Methods for Scientists and Engineers* by R.W. Hamming. Copyright © 1962 by the McGraw-Hill Book Company, Inc. Used with permission of McGraw-Hill Book Company.

"carry out the list of instructions" can be given, and you are on your way. Even for tasks much more complicated than adding a long list of numbers, you will usually have your answer in a matter of seconds.

The list of instructions that tells the computer how to accomplish a certain task is called a *program*; and the process of (1) analyzing the task as a sequence of basic steps to be performed in a specified order, (2) translating this analysis into a language the computer can understand, and (3) communicating the resulting program to the computer is called *programming* the computer. The most important feature distinguishing a computer from other computational devices—such as, abacus, slide rule, and desk calculator—is the stored-program concept that makes it possible to take advantage of the lightning speed of the electronic components that actually do the arithmetic. (We can program other computational devices as well, but we keep the programs in our heads or on paper and exercise direct, step-by-step control of the devices.)

Internally, a computer processes only *numerical* information, and that is usually expressed in a *binary* code using only 0 and 1 as digits. The instructions embodied in a program must ultimately be expressed in this code for the computer to understand them. Learning to program a computer in its own language is a long and tedious job, but fortunately other programmers have been there ahead of us and have provided built-in programs called *interpreters* and *compilers* that will accept instructions written in other languages and translate them into instructions the computer can understand. This process is invisible to most users, and the illusion will be created that you are communicating with the computer in a language that is very similar to your own when you talk about mathematics in English. There are quite a few sophisticated programming languages available, but for problems of relatively low complexity such as the ones we consider, all the popular languages look pretty much alike.

In this book we use the most popular language for instructional computing, BASIC, as the vehicle for discussing programming. If you have not used a computer before, Chapters 1, 2, 4, and 5 introduce enough about BASIC to prepare you for the exercises in the rest of the book. If you already know BASIC, please skim through this material, but feel free to skip anything you find too trivial.

If you are studying or already know another language (e.g., Pascal or FORTRAN), you are encouraged to use it instead, and you should not find the translations difficult. Beyond Chapter 6 the text is almost entirely language-independent.

0.3 MODES OF COMPUTER USE

Your computer may be a *micro*computer, a *mini*computer, or a *maxi*computer. (The last is usually called a "mainframe" computer.) The implied size differences are not important for anything we do in this book, but each type of computer has some characteristic modes of use, and variations in these may occasionally force us into brief digressions to serve the needs of those with a system that is different from your own.

A microcomputer is a small enough device to sit on a desk. It typically serves just one user at a time, and that user has total control over the computing environment, including the off–on switch and the choices of peripheral devices and software (programs) to use. Micros are also sometimes "clustered" or "networked" in multi-user configurations that may include shared access to a storage device for programs (e.g., a "hard disk" drive), a printer, and/or communications to a larger computer. In such configurations the individual user still has total control over the functions of his or her micro, but not over the shared facilities. A typical microcomputer system includes a keyboard with a typewriterlike layout of keys, a video display (TV set or monitor), and a mass storage device (cassette tape drive, floppy disk drive, or hard disk drive). It may or may not have a printer attached for "hard-copy" output. Video displays for micros can vary from as little as a few lines of 20 characters each (rare and barely usable) to 40 lines of 80 characters each (ideal, but expensive). We assume a "typical" display of 24 lines with 40 characters each, and when that is too limiting for the problem at hand, we say so. The nature of the storage device is not important, as long as there is one. We leave it to your instructor to explain how it works. We do not assume that a printer is available; in the few places where one is needed, that is spelled out.

A minicomputer typically occupies as much space as one or several pieces of office furniture (desks and filing cabinets). A mainframe computer may occupy one or several large rooms. The distinction between these two types of computer is not important for the individual user, since these facilities are usually shared by many users. A mini may serve anywhere from a single department or laboratory to an entire (small) campus, and a mainframe may serve a whole campus or a whole state or regional network of campuses. Users communicate with such computers via "terminals" in one of two modes: (1) "conversational" or "interactive" mode; or (2) "batch" or "remote" mode. In the first of these modes, the computer itself is invariably operating in a "time-sharing" mode, sharing its resources with many users at once, and this may be true even if the communication is taking place in batch mode.

Time-sharing was invented, before microcomputers became a reality, to give the small user of a large system the illusion of total control over the computing environment, without the obvious waste of resources that would occur if the large computer were frequently idle, waiting for the user to decide what to do next. Thus the user of a time-sharing system via an interactive terminal may see little difference from using a microcomputer, except there are a few more steps involved in getting started, and response time of the system may be a little slower if it is heavily loaded. Communication with a batch system is typically done with punched cards prepared in advance on a keypunch machine, but some systems allow interactive preparation of electronic "card decks" that are submitted to the batch input stream of the computer without physical punching or reading of cards.

Regardless of whether input to the computer is being accomplished via a video terminal, a printing terminal, or punched cards, the user communicates his or her wishes with a typewriterlike keyboard, so we refer to the input device as the *keyboard* throughout this book. Output devices for time-sharing and batch systems are usually less restricted than those for microcomputers, except in one important respect: Many of the popular micros provide graphics capabilities that are lacking on most terminals and printers attached to larger systems. We deal with this distinction more fully in Chapter 6.

0.4 A BRIEF GLOSSARY

Some or all of the following technical terms may be unfamiliar to you. Some of them have already been used in this chapter; others will appear later in the text.

BATCH PROCESSING Grouping several similar programs to be run sequentially on the computer. (This process is usually not visible to the user, but it is the normal processing mode when programs and data are being entered via punched cards.)

COMPILER A program (usually provided with the computer system) that translates programs written in a user-oriented language, e.g., Pascal or BASIC, into a machine language program.

CONVERSATIONAL COMPUTING Direct-access use of a computer in which messages are entered via a typewriterlike device and responses from the computer are returned, usually very quickly, to the same device. The user has the impression of carrying on a conversation with the computer. Also called "interactive computing" (see Sect. 0.3).

DEBUGGING The process of locating and correcting errors in a program (see Chap. 5).

DISPLAY As a noun, a cathode ray tube (CRT) device, such as a TV set or video monitor, on which computer output and/or input are displayed.

DOUBLE-PRECISION ARITHMETIC Arithmetic using the equivalent of (up to) 14 to 16 significant decimal digits in all steps. Compare single precision (see Chap. 13).

DIRECT INQUIRY A capability of some conversational computer systems of asking the computer questions—such as, "what is four times the angle whose tangent is 1," "what time is it," "what is the last known value of X in the program just completed?" Sometimes called "desk-calculator mode."

EXECUTE Carry out the sequence of instructions represented by a program. Synonym: Run.

FILE A linear arrangement of data and/or instructions on a device suitable for computer processing. (Think of the individual folders in a filing cabinet drawer or file cards in a box.) A computer file may be on an electromagnetic device, such as a disk or tape, or it may be on punched cards or punched paper tape. One normally creates a file in the process of preparing a program to be run by the computer.

FLOWCHART A diagrammatic representation of the logical sequence of steps that are necessary to solve a problem. A useful tool for building a computer program (see Chap. 1).

INPUT As a verb, to enter information to a computer from an external device, such as a keyboard, a disk, a magnetic tape, or a deck of cards. As a noun, that which is input (e.g., a program, or the data to be processed).

INTERACTIVE COMPUTING See conversational computing.

INTERFACE A common boundary between two devices, systems, or realms of thought.

INTERPRETER A program that decodes instructions written in a user-oriented language and then immediately executes those instructions, in contrast to a compiler, which creates a machine language file to be executed as a unit as a separate step. The distinction is usually invisible to the user.

KEYBOARD An input device resembling a typewriter keyboard (usually with QWERTY . . . arrangement of keys). With minor variations, common to microcomputers, interactive terminals, and keypunch machines (see Sect. 0.3).

LOCATION A physical unit within the computer containing a unit of information, such as the value of a certain variable.

OUTPUT   As a verb, to transmit information from a computer to an external device, such as a printer, video display, magnetic tape or disk, or a punched card. As a noun, that which is output (e.g., answers).

PROCESSOR

1. (Software) A program provided as part of a computer system to process—i.e., translate and perhaps run—programs written in a certain language. Processors include both compilers and interpreters.

2. (Hardware) The electronic circuitry that carries out the processes for which a computer is designed, e.g., arithmetic. One of the developments that made microcomputers possible was the invention of very small processors called *microprocessors*.

PROGRAM   As a noun, a list of instructions to a computer to accomplish a specific task. As a verb, to create such a list (see Sect. 0.2).

RUN   See Execute.

SINGLE-PRECISION ARITHMETIC   Arithmetic using the equivalent of (up to) seven or eight significant decimal digits in all steps. Compare double precision (see Chap. 13).

TERMINAL   An input/output device that is capable of sending information to and receiving information from a computer. An interactive terminal usually consists of a keyboard and either a display or a printer. A batch terminal usually consists of a card reader and a printer. A terminal that is not in the same room as the computer is called a *remote terminal*.

References

Boyer (1969); Cajori (1919); Cohen; Gardner (1970); Goldstine; Kemeny; Larrivee; Morrison and Morrison; Rosenthal; Schrader; Spencer; Ulam.

C H A P T E R 1

COMPOUND INTEREST: FIRST STEPS IN PROGRAMMING

In this chapter we analyze two variations of a very elementary programming problem by using flowcharts. Then we show how the flowcharts can be translated into a programming language that is understood by most computers. The procedures used for analysis and programming are useful in more complicated situations later, and the programs are employed for illustrative purposes in your first session with the computer. The language used is called BASIC, which stands for Beginner's All-purpose Symbolic Instruction Code. It is a student-oriented language that was designed originally for use with conversational terminals to time-sharing computer systems and is now by far the most popular language for programming microcomputers. It is a language with many different dialects, but the features we employ are common to almost all of them. That means we forego use of any of the clever additions to BASIC that may be part of your particular system, but you are not restrained from learning about them and using them if you wish.

1.1 HOW DOES YOUR MONEY GROW?

Problem 1 Suppose a bank is offering three-month "money market" certificates at 8.8% annual interest with the option of reinvesting the principal and interest every three months, i.e., of compounding the interest quarterly. If you were to hold such certificates for 20 years, how much would your initial investment have grown over that period?

Note that the initial deposit is not specified, so let's denote it by a variable, P (for *principal*). Each quarter the principal P will earn interest $I = (0.022) P$. (Why?) The new value of the principal will then be $P + I$. Thus to solve our problem, we need only repeat 80 times the quarterly computation of interest followed by the increase in principal. Prior to the start of this repetitive computation, we need to provide a value of P to the computer, and after completion of the 80 steps, we want the computer to tell us the final value of P. We summarize the discussion of the problem in the *flowchart* shown in Figure 1.1. Such diagrams will become increasingly useful as we consider more complicated problems. What we have done so far is to break up our total task (find the principal 20 years after deposit) into steps that we arrange in a logical sequence. Each box in the diagram represents a specific task, which corresponds roughly to a programming step. How specific a task needs to be is something you will learn as you go along. The central part of Figure 1.1 is an *indexed box,* or *loop,* which contains the specific tasks that are to be repeated 80 times. The indexing information $N = 1, 2, 3, \ldots, 80$ indicates that the loop is indexed by N (a variable introduced for the purpose of counting quarters), and that N is to take each integer value from 1 to 80. This means that the individual steps in the loop are to be carried out *first* with $N = 1,$ *then* with $N = 2,$ and so on, *finally* with $N = 80.$

Figure 1.1

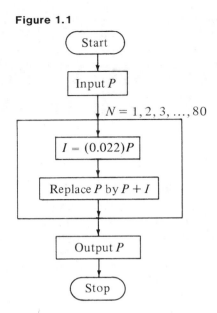

Here is a program that will carry out the steps indicated in the flowchart:

```
10 REM   PROGRAM 1.1
20 REM   COMPOUND INTEREST FOR 20 YEARS
25 REM   AT 8.8 PERCENT
30 INPUT P
40 FOR N=1 TO 80
50 LET I=.022*P
60 LET P=P+I
70 NEXT N
80 PRINT P
90 END
```

Many important features of the BASIC language are illustrated in Program 1.1, and we comment on these in detail in the order in which they appear (not in order of importance).

1. A program in BASIC consists of a sequence of numbered statements, each of which can be typed on one line (no subscripts or superscripts) on a keyboard. The numbering chosen is quite arbitrary, except that the computer will consider the statement (when carrying out the instructions) in numerical order. The gaps of 10 are left to make it easy to insert integer-numbered statements later (corrections or additions).

2. The first three lines of the program are *remarks,* placed there for easy identification of the program by a human reader, and ignored by the computer when it carries out the instructions. Such identification is very useful for keeping track of what you are doing, and your instructor will tell you ways to use comments so that he or she can keep track of what you are doing. Any text that can be typed on your keyboard can be placed after REM as a remark.

3. Line 30 directs the computer to INPUT a value for the variable *P*. During the running (or execution) of the program (i.e., carrying out the instructions by the computer), this instruction causes the computer to pause and wait for you to enter a numerical value from your keyboard. That value is then assigned to *P*, and execution resumes with the next numbered statement. Thus it is not necessary to decide now what *P* will be. Furthermore, it is very easy (as we will see) to execute the program as many times as we like with a different value of *P* each time.

4. Statements 40 and 70 are a FOR-NEXT pair, represented in the flowchart by the indexed box. The effect of this pair is to cause the execution of all the statements between the two, once for each value of the index *N* specified by the condition following FOR. When, as in this case, only the first and last values of *N* are specified, the increment in *N* is assumed to be 1; i.e., *N* will take the values 1, 2, 3, 4, . . . , 79, 80, and therefore the statements numbered 50 and 60 will be executed exactly 80 times.

5. Statement 50 is an *assignment,* indicated by the word LET, followed by a variable name, followed by an "equals" sign, followed by some sort of expression. The effect of an assignment statement is to compute the value of the expression on the right and to assign that computed value to the variable name on the left.

6. The asterisk(*) in statement 50 stands for multiplication. Thus the expression to be computed is just the product of the *constant* 0.022 and the value of the variable *P* (which will be different each time through the loop, because of statement 60, but one thing at a time). It is necessary to have a multiplication symbol rather than just writing, say, *AB* for the product of *A* and *B*. The usual · (raised dot) and × (cross, not the letter x) are not usually available as keyboard characters, so * is used.

NOTE: The use of the value of *P* in line 50 does not change that value or erase the contents of the corresponding storage location.

7. Statement 60 is also an assignment statement, but the use of the "equals" sign may look a little curious here, since *P* will never actually equal *P* + *I*. However, the definition of assignment given earlier tells us what this statement also does. The expression on the right is computed first, i.e., the value of *P* (current principal) is added to the value of *I* (newly computed interest), and the sum (new principal) is assigned as the value of *P*. Assignment of a *new* value to a variable that already had a value wipes out the old value, so this statement has exactly the effect that is indicated in the flowchart: it replaces the old value of *P* by *P* + *I*, which becomes the new value of *P*.

NOTE: As before, even though the value of *I* is used in evaluating the expression on the right, that value is not changed in storage.

8. Statement 80 is an *output* statement that directs the computer to PRINT the current value of the variable *P*, i.e., to tell us the answer. Our flowchart shows that this step is reached only after the loop has been completed, so we will see only one value of *P*, the principal after 20 years.

9. Line 90 serves two functions: (1) It stops the running of the program. (2) It identifies the END of the program (physically the last line) to the BASIC processor. A STOP statement will do as well for the first function, but some processors require that the last

line of every program (and only that line) be an END. Most microcomputer BASICs *do not* have this requirement, and in fact treat END and STOP as synonyms.

10. We have used *P*, *I*, and *N* as variable names in the program (and in the flowchart). Any single letter, or a letter followed by one digit (0 through 9), is a legitimate variable name. For example, *A*, *Q*, *X*, *Z*7, *M*9, and *B*2 are all proper names for variables.

Now that you have read comments (1) through (10), turn back to Figure 1.1 and Program 1.1 again. Read the program carefully, and compare it with the flowchart. Be sure you understand each step, as well as how the steps fit together to accomplish the desired goal. If any part of the programming process for this problem remains confusing, study the appropriate comments again. If that doesn't help, take it up with your instructor as soon as possible because understanding at this point is crucial for future success with programming.

When you are satisfied that you understand the program, go on to Section 1.2 to read about a variation of the original problem that will introduce some additional important aspects of programming.

1.2 HOW LONG DOES IT TAKE TO DOUBLE YOUR MONEY?

Problem 2   Suppose we have the same situation as in the previous section, namely, savings certificates paying 8.8% interest compounded quarterly. How long will it take to double the initial investment?

As a first step in analyzing this problem, we might observe that the principal must surely double in *less than* 20 years Why? (HINT: How long would it take to double your money at only 5% *simple* interest?) Thus at some point during the execution of the loop in Program 1.1 the new *P* must be at least twice as large as the original investment. At that point the value of the variable *N* will be the elapsed time in quarters, so *N*/4 will be the elapsed time in years. Hence our problem can be solved by inserting into the loop a decision step that interrupts the loop at the appropriate time and tells us what *N*/4 is. Note that, in Program 1.1, the initial value of *P*—i.e., the value provided through the input statement—is lost after the first quarter. To have the initial value available for comparison, we must make a copy of it that will not be changed during the computation. This is done by making it the value of another variable, say, *Q*, before the process of increasing *P* begins. A flowchart summarizing this analysis is shown in Figure 1.2.

The most important new feature of this diagram is the *decision* block, the one with the diamond shape. The contents of such a block normally is a question that can be answered *yes* or *no;* and there are separate arrows leaving the diamond, *appropriately labeled,* to indicate what the next step is in each case. Here, a "yes" answer leads to termination of the loop and output of the desired answer, *N*/4. The "no" arrow leads to the bottom of the loop, which is the signal to add 1 to *N* and begin again at the top of the loop (unless *N* = 80, of course, which should be impossible).

The flowchart also illustrates two important features of the loop notation. First, interruptions of loops are indicated by an exit through a *side* of the box rather than through the bottom. Second, there must always be *exactly one* arrow from the bottom of the box to indicate what to do next after the loop is completed. This is important even if you think

Figure 1.2

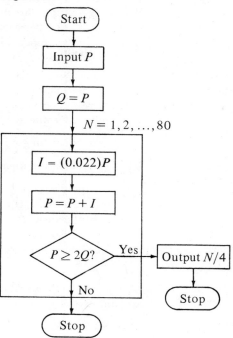

completing the loop is impossible, because a mistake in analysis, programming, or typing might lead to unexpected completion of the loop.

Here is a BASIC program to implement the flowchart:

```
10 REM   PROGRAM 1.2
20 REM   TIME-TO-DOUBLING PROGRAM
30 INPUT P
35 LET Q=P
40 FOR N=1 TO 80
50 LET I=.022*P
60 LET P=P+I
65 IF P>=2*Q THEN 100
70 NEXT N
80 PRINT "ERROR"
90 STOP
100 PRINT "TIME-TO-DOUBLING IS ";N/4;" YEARS"
110 END
```

The following series of comments explain the new features of the language.

1. There are two important new ideas in line 65 of the new program. First consider the following portion of that line:

$$P >= 2 * Q$$

This is the statement that P is greater than or equal to two times the initial investment, a statement that is either true or false depending on the value of P. It corresponds directly to the contents of the decision block in Figure 1.2. The symbols $>$ (greater than), $<$ (less than), and $=$ (equals) are used for such relational statements in the same way that they are used in mathematics. However, there are no single symbols for $\geq$ (greater than or equals) or $\leq$ (less than or equals), so the compound symbols $>=$ and $<=$ are used. The quantities to the left and to the right of such a relational symbol must be expressions that the computer can evaluate in terms of current values of the variables involved, so that the truth or falseness of the entire statement can be determined.

2. Now consider the entirety of line 65, which is of the form

$$\text{IF} \ \langle\text{relation}\rangle \ \text{THEN} \ \langle\text{statement number}\rangle$$

The purpose of this statement is to carry out the decision step in the flowchart. IF the relation ($P \geq 2Q$) is true, we want to interrupt the normal sequence of taking instructions in numerical order, exit from the loop, and print the answer. THEN 100 transfers control to line 100 in this case. However, if the relation is false, no transfer takes place and control passes to line 70 (the bottom of the loop). Thus line 65 does indeed carry out the decision block *and* both of the arrows leaving it.

3. In statement 100 I have added a slightly fancy (but unnecessary) touch, namely, the addition of text to the output. This line could just as well have been

$$100 \qquad \text{PRINT} \qquad N/4$$

If the answer turned out to be, say, 15.25 years (which happens not to be the correct answer, as we will see), then execution of the simpler version would cause the terminal to print:

$$15.25$$

On the other hand, the version of line 100 in Program 1.2 will cause the terminal to print:

$$\text{TIME-TO-DOUBLING IS 15.25 YEARS}$$

In problems with several answers, such identification of output is very useful in keeping track of what's happening. Any characters that are available at your terminal may be enclosed between quotation marks; the result is called a *character string*. When a character string appears in an output statement, the effect is to cause the printing of the text exactly as it appears in the string. The effect of the semicolons separating the character strings from the numeric output is to print the three items without spaces between them. (Note the spaces within the strings themselves.) Output items may also be separated by commas, but the effect of commas varies among different dialects of BASIC. The usual effect is to place numeric output items in predetermined columns.

4. Note that the numerical expression $N/4$ in line 100 is *not* the name of a variable, but rather, must be computed from the variable N. Any arithmetic expression (such as those discussed earlier as right-hand members of assignment statements or either member of a relational statement) may be an entry in an output list.

5. Note that division (as in $N/4$) is indicated by a slash, rather than a horizontal line, for the obvious reason: The text of a given expression has to be typed on a single line. In more complicated expressions it is often necessary to use parentheses to keep numerators and denominators together.

6. This program has two possible stopping places, lines 90 and 110. There is a good reason for lines 80 and 90. If they were not there and a mistake caused completion of the FOR-NEXT loop, then control would pass next to line 100, causing the printing of what appeared to be an answer, but it would be wrong. On the other hand, execution of lines 80 and 90 causes the printing of the word ERROR and a STOP before line 100 is encountered, so there can be no doubt that something went wrong.

7. There was a reason for our choice of line numbers in Program 1.2. Lines 30, 40, 50, 60, and 70 also appear in Program 1.1. Anticipating that we take these programs to the terminal at the same time, we do not need to retype these lines when switching from Program 1.1 to 1.2 (as we will see).

As you did with the first problem, you should now turn back to Figure 1.2 and Program 1.2 and review them step-by-step. Be sure you understand each block in the flowchart, each statement in the program, the correspondence between flowchart and program, and how it all fits together to accomplish the desired task.

When you are satisfied that you understand the program, you are ready to turn to Chapter 2. If, however, you would like to try your hand at writing your own programs for some more general compound-interest problems, you may go on instead to (optional) Section 1.4.

The next section summarizes the features of BASIC that have appeared in our sample programs and also introduces one more important statement (GO TO) and one more operation (exponentiation) that has not occurred as yet.

1.3 SUMMARY AND MISCELLANEOUS COMMENTS

We summarize here the most important features of BASIC, important in the sense that they will be adequate for all the programming required in studying this book.

1. *Constants* in BASIC may be written just as you write them in mathematics, e.g., 0, -1, 17.25, 3.1416, -10000 (but don't insert commas). They may also be written in "scientific" or "floating decimal" notation when convenient. For example, 5×10^8 can be written 5E8 or .5E9. (Read the E as "times 10 to the power.") Similarly, -0.000003 can be written $-3E-6$ or $-.3E-5$ or $-300E-8$. Answers that are very large or very small may be printed by the computer in scientific notation.

2. *Variable names* in BASIC consist of a single letter, or a letter followed by a single digit. Note that the computer does not see any connection between the names A, $A0$, $A1$, $A2$, and so on. They just refer to distinct storage locations, as would the names A, J, P, X. (Some versions of BASIC allow more general names for variables.)

3. *Assignment* statements have the form

$$\text{LET}\langle\text{variable}\rangle = \langle\text{expression}\rangle$$

where the expression on the right may be formed from variables, constants, parentheses, and arithmetic operations $(+, -, *, /)$, as you would with mathematical expressions. Later we see that you may also use certain functions in forming expressions, some of which are built into your system, and some of which you define yourself. (Many versions of BASIC allow "LET" to be omitted in assignment statements.)

4. There is one additional arithmetic operation that did not appear in Chapter 1 but which is needed later: exponentiation, denoted by $\uparrow$. Thus A $\uparrow$ B, where A and B represent any expressions, stands for A^B. (Some systems use $\hat{\ }$ or $**$ for exponentiation. Be sure to check which form is right for your system.)

5. When the order in which arithmetic operations are to be performed is not forced by the use of parentheses, that order is determined as follows by a *hierarchy* of operations: Exponentiations are performed first, then multiplications and divisions, then additions and subtractions. Within levels of this hierarchy (e.g., with two multiplications, or with an addition and a subtraction), the operations are performed in left-to-right order. Check the examples in Table 1.1 carefully.

6. *Loops* are formed by FOR-NEXT pairs of statements, which may be thought of as corresponding to the top and bottom of an indexed loop box in a flowchart. The general

Table 1.1

BASIC	Interpretation
A * B $\uparrow$ C	$A \cdot B^C$
(A * B) $\uparrow$ C	$(AB)^C$
A $\uparrow$ B * C	$A^B C$
A $\uparrow$ (B * C)	A^{BC}
A * (B − C)	$A(B-C)$
A * B − C	$AB - C$
A/B − C	$\dfrac{A}{B} - C$
A/B * C	$\dfrac{A}{B} \cdot C$
A/(B * C)	$\dfrac{A}{BC}$
A − B + C	$(A - B) + C$
A − (B + C)	$A - (B + C)$
A $\uparrow$ B $\uparrow$ C	$(A^B)^C$
A $\uparrow$ (B $\uparrow$ C)	A^{B^C}

form of the FOR statement is

$$FOR\langle variable\rangle = m_1 \text{ TO } m_2 \text{ STEP } m_3$$

where "variable" may be any variable name (the index variable), and m_1, m_2, and m_3 may be any constants, variables, or expressions. The statements inside the loop are first executed with the index equal to m_1, then with $m_1 + m_3$, $m_1 + 2m_3$, and so on, until the index exceeds m_2, at which time control passes to the next statement after the loop. The NEXT statement has the form

$$NEXT\langle variable\rangle$$

and the variable must match the index variable of the FOR that it goes with.

NOTE: If "STEP m_3" is omitted, m_3 is assumed to be 1.

7. *Conditional transfers* of control are accomplished with

$$IF \langle relation\rangle \text{ THEN } \langle statement\ number\rangle$$

where the relation is formed by putting a relational symbol (e.g., $>$, $<=$, or $=$) between two expressions. If the relation is true, control passes to the statement whose number is specified. If false, the normal sequential order is continued.

NOTE: When you are writing a program, you often find that you don't know the statement number to which you want to transfer, because it comes later in the program and you haven't written it yet. Just leave it blank, and go back and fill it in later, after you have written the rest of the program. Your flowchart will help you identify the correct transfer points.

8. *Unconditional* transfers of control are also possible, and sometimes necessary, even though they did not turn up in Programs 1.1 and 1.2. The proper form for such a statement is

$$GO\ TO \langle statement\ number\rangle$$

For example, suppose we want to set X equal to 7 if A is positive, but set X equal to $Y + Z$ otherwise, and in no case to do both. A program segment to do this might look like

```
      ...
 90   IF A > 0 THEN 120
100   LET X = Y + Z
110   GO TO 130
120   LET X = 7
130   ...
```

NOTE: GO TO statements correspond to *arrows* in flowcharts, not to blocks. Most arrows do not need GO TO's, however, since they represent the normal sequential order of the program.

9. *Input* is accomplished with INPUT, followed by one or more variable names, separated by commas (if there is more than one). For example,

$$INPUT\ A, B, C$$

causes the computer to print ? and wait for you to respond with a value for A, then one for B, then one for C. You may enter these one at a time on separate lines, or all at once, separated by commas. If you have a lot of INPUT points in your program, it's a good idea to precede each INPUT by a PRINT statement (with character string output) to print the name(s) of the variable(s) that are to be entered.

10. *Output* is accomplished by PRINT, followed by one or more variables, expressions, character strings, or constants, separated by commas or semicolons. Semicolons cause items to be printed adjacent to each other. Commas cause them to be printed in specified fields, which vary from one system to another. These fields may be from 10 to 15 characters wide, and there may be from two to five of them per line.

11. *Remarks* are indicated by REM at the beginning of a line. Any such line is ignored by the computer when the program is being processed.

12. *Spacing* items on each line is entirely up to you, with certain obvious exceptions: You must not space in the middle of a word (e.g., IN PUT instead of INPUT), and you must leave at least one space between distinct items (e.g., NEXT N, not NEXTN). Spacing should be used judiciously to improve readability. For example, it is common practice to indent the statements between a FOR-NEXT pair, which makes it easier to see where the loops are and which pairs go together. (Some systems *do not* require spaces between distinct items, but you should use them anyway. Some systems mess up your clever indentations and reformat lines their own way after you enter them. And some expect GOTO to be written as one word instead of two.)

1.4 WHAT TO DO IF YOU ARE BORED [O][1]

If you have already studied computer programming, the problems posed in Sections 1.1 and 1.2 are probably too simple to hold your interest. (No pun intended.) Here are some exercises that you can work on while waiting for your classmates to catch up.

Exercises

1. You wish to deposit a certain sum P in a savings account that pays X percent per annum, compounded K times a year. How much will you have in the bank after Y years? P, X, K, and Y are to be treated as input variables. Make a flowchart, similar to Figure 1.1, to express the logical order of steps for solving this problem, and write a program to carry out the computation.

2. Given the situation of Exercise 1, find the time required for doubling the initial investment—i.e., plan and write a program to find the time.

3. Make a flowchart and corresponding program to solve the two preceding problems simultaneously. (This is harder than it looks.)

4. If you have been thinking a little bit about what was going on mathematically in Section 1.1, you may have noticed that the loop is not really necessary for solving this very simple

---

[1]The symbol [O] designates material that is optional and intended only for those students who are prepared to go beyond the basic parts of the text.

problem. With a little algebra you can figure out a formula for the principal after 20 years in terms of the initial principal. Program 1.1 may then be replaced by a four-line program: input, formula, output, and stop. Write such a program.

5. Repeat Exercise 4 in the more general context of Exercise 1.

NOTE: The time-to-doubling problem can also be solved by a simple formula by using logarithms. We return to this subject in Chapter 41.

References

Kemeny and Kurtz; Rosenblatt and Rosenblatt. You should also have access to a BASIC user's manual appropriate for your computer system.

CHAPTER 2

GETTING YOUR HANDS ON THE COMPUTER [L]

This chapter is the first of a number of "lab manual" chapters (those designated [L]), intended to guide you as you work at a computer or terminal. The objective for your first session with the computer is to see that the programs prepared for you in conjunction with Chapter 1 actually produce the desired answers. This will give you some confidence that the machine actually works the way we have been saying it would. If you have prepared any programs in connection with the optional exercises in Section 1.4, those programs may be substituted for ours, but we make no further reference to that section.

2.0 PRELIMINARIES

This chapter necessarily begins with a large, gaping hole. Because of the wide variety of hardware and software (system programming) provided for interactive computing, we cannot provide you with a detailed list of instructions for the procedures to follow in using your particular computer or terminal. However, we can provide you with a checklist of the important tasks you have to know how to perform. Your instructor will have to fill in the details, preferably with a written handout. If those instructions to you miss one or more of the following points, ask about it. You need to know how to

1. Turn on the computer (if you are using a micro) or call up the computer (if you are using a terminal to a time-sharing system);

2. Provide proper identifying and accounting information; i.e., log in (usually not necessary with micros);

3. Gain access to the appropriate language processor (some micros have BASIC active when turned on);

4. Create and name a file (included in step 8 on some systems);

5. Number lines in your file;

6. Enter complete typed lines (usually ENTER or RETURN key);

7. List your program (LIST command in BASIC);

8. Save your program (SAVE command in BASIC);

9. Run (or execute) your program (RUN command in BASIC);

10. Log out when you are finished (terminal) or turn everything off (micro).

In addition, and just as important, you need to know how to make simple corrections and changes, when the instructions given to the computer happen not to be exactly what you had in mind—e.g., as the result of a typing mistake. This includes knowing how to

11. Cancel an incorrect character in a line before it is entered (usually "backspace," but not all keyboards have such a key);

12. Cancel a whole line before it is entered;

13. Erase or delete a line after it is entered (in BASIC enter just the line number again);

14. Change a line after it is entered (in BASIC entry of a new line erases the old one with the same number);

15. Insert an additional line after the program has been entered (in BASIC just use an intermediate line number); and

16. Get the computer's attention when it isn't doing what you want it to do (e.g., BREAK, ATTN, RESET, or RUN/STOP keys).

Eventually you will need to know how to

17. Load a saved program from disk, tape, or library file (LOAD command in BASIC);

18. Delete (scratch, erase, "unsave") a saved program; and

19. Replace a saved program with one that has been changed by one or more of steps 11 through 15 (on many systems you just save again with the same name).

Your system may have some additional useful features that your instructor will want to tell you about, but the preceding list contains all the basic ones and is sufficient for carrying out all the computer exercises in this book.

As noted in Section 0.3, your communication with the computer takes place via a device with a typewriterlike keyboard. When you first sit down at the keyboard, be sure to note the locations of the special keys associated with functions 2, 6, 11, 12, and 16. Also, take note of the locations of characters that are not standard on a typewriter—e.g., the relational operators > and < that are used in BASIC. Notice too that the digits 0 and 1 have their own keys, and you must not use capital "oh" and lowercase "el" for these.

It is helpful if you already know touch typing. If you don't, you may feel intimidated by the keyboard. However, typing skill is not essential; "hunt and peck" is adequate for getting the job done, and you will quickly find yourself getting faster at it.

One last word of comfort before we actually begin: Don't be afraid to push buttons or keys. Your system has been designed so that you can't hurt it. The computer may complain from time to time when you do something unexpected, but you can't make fuses blow or important programs vanish or do anything worse than prematurely breaking your connection with the computer. (If that happens, start again.)

2.1 RUNNING THE PROGRAMS

Now that you have gotten to know your computer, you are ready to get to work. We suppose that you have started up or logged in to your system and are ready to type in Program 1.1, the program for solving the 20-year compound-interest problem. You do this by creating a file whose numbered lines are the statements in your program. When you have finished typing the program, you do what is necessary to signal the end of the file, and you may want to (or have to, depending on your system) save it. Then you do what is needed to execute it; in other words, you have to invoke the appropriate processor residing in the inner recesses of your computer and feed it your completed file (program).

Turn back to Chapter 1 now, and type in Program 1.1. (If you are using a language other than BASIC, your instructor will provide a substitute program.) If you make a typing mistake, refer to your instructions corresponding to items 11, 12, 13, or 14 in Section 2.0. When you have finished typing the program, LIST it. Check carefully to see that it looks exactly like the program in the text. (If not, make additional corrections.)

When you are satisfied that the program is correct, SAVE it, and then RUN it. (When you run or execute a program, you are ordering the computer to carry out your instructions, one at a time, in numerical order.) The first three lines of your program are skipped because they contain only REMarks, and the first real instruction is to acquire a value for the variable P. Thus almost immediately the computer indicates that you are now to enter an appropriate initial principal. Try typing in

$$2000$$

and remember to "enter" your input (i.e., press the RETURN or ENTER key). Once you have assigned a value to P, the rest of the steps in the program will be executed very quickly, and an answer will appear immediately below your 2000. There will probably also be an indication that the END statement was reached.

Want to see that again? Execute again, and provide a value of P of your own choosing. You may do this as many times as you like.

Now we are ready to work on Program 1.2, which you will find in Section 1.2 (or a handout from your instructor if you are not using BASIC). Notice that lines 30, 40, 50, 60, and 70 are the same as they were in Program 1.1. Type in lines 10, 20, 35, 65, 90, 100, and 110 as they appear in Program 1.2. New lines with old line numbers replace the statements previously on those lines. Lines with new numbers are automatically inserted in the proper places in your program. Now LIST your program; it should look just like Program 1.2.

If you are satisfied that the program looks right, it is time to SAVE it (with a different name from that used for Program 1.1) and RUN it. Start with $P = 2000$, as before. Also RUN again (more than once if you like) with another value of P. Are you surprised by the answers, or are they what you expected?

2.2 POSTSCRIPT

If you have looked closely at the answers from several runs of Program 1.1, you may have observed that the ratio of the final balance to the initial investment is (apparently) constant. Let's see why this is so. The effect of compounding the interest ($0.022 \times P$) and adding this amount to P is to replace the old value of P by $P + 0.022 \, P$, or equivalently, by $1.022 \times P$. Thus each trip through the loop has the effect of multiplying the principal by the constant factor 1.022. When the loop is carried out 80 times, the initial investment is multiplied by this factor 80 times, so the final balance must be

$$(1.022)^{80} \text{ times the initial deposit.}$$

Now you have a chance to create your own program to see if this simple formula gives the same answer as Program 1.1. First, clear the lines of Program 1.2 (or create a "new" file); the command is NEW in BASIC. Now prepare a two-line program that will PRINT

$$2000 * (1.022 \uparrow 80)$$

and END. RUN your program to see if you get the same answer as that given by Program 1.1 with $P = 2000$. (If your system permits a "direct inquiry" or has a "calculator mode," you can get the value of the preceding expression without writing and running a "program." Check with your instructor about this. Of course, your pocket calculator will do just as well at this task.)

Finally, recall your instructions for what to do when you finish a laboratory session (item 10 in Sect. 2.0); this session is now finished.

C H A P T E R 3

ALGORITHMS: AN ETYMOLOGICAL INTERLUDE [R][1]

Once you have been exposed to the tremendous power that is at your fingertips with a high-speed computer, there is some danger that you will become so fascinated by the tricks you can make it do that you will forget to eat, sleep, or go to class. Solving a little problem in compound interest may not be enough of an initiation into the mysteries to justify that statement, but before long you will see your friend do some truly marvelous things. (And at times of frustration when you can't figure out what it's doing, you may see it as an insidious adversary rather than as a friend.)

But we must not lose sight of the forest as we plunge into the trees. The important part of what we are doing with the computer is *solving problems.* Apart from the details of learning a new skill, the important lesson of Chapters 1 and 2 is that problem solving (with a computer, and often with other devices as well) proceeds in the following manner. The problem must be analyzed to reduce it to essential details that are expressed mathematically. A method for solution must be formulated and expressed step-by-step in terms of very simple operations. Finally, the step-by-step procedure must be communicated accurately to the device (in this case, the computer) that is to carry it out.

A step-by-step procedure for solving a problem is called an *algorithm.* The analysis phase of problem solving is not a new idea to you, for you have surely seen it in other mathematics courses. The final phase of programming and communicating with the computer may require some practice (as does any newly acquired skill), but it is not an intellectually deep activity. The important connecting link is the middle phase, the formulation of an algorithm.

Algorithmic thinking may be a new, and sometimes challenging, intellectual activity for you. Often in previous mathematics courses you have taken and even in the calculus course in which you are now engaged, solution methods may be presented as one-step formulas—e.g., the quadratic formula; but how do you take a square root without help? Or attention may be focused primarily on specific *steps,* rather than on the integration of these steps into an entire procedure that is adequate for solving a really complicated problem (the trees again, rather than the forest).

Every student of the English language (including, perhaps reluctantly, every graduate of an American high school) knows that words become part of our language in some strange ways. But the word "algorithm" has a more curious origin than most.

Our story begins with the expansion of the Muslim Empire during the century A.D. 650 to 750. The center of this empire shifted from time to time, depending on who was in charge. At the peak of expansion, when the empire extended from Spain to the edge of China, the

---

[1]Material designated [R] is to be read by students but not to be taught or tested.

Abbasids, a dynasty descended from Mohammed's Uncle Abbas, massacred the ruling family in Damascus and founded a new capital at Baghdad. This move coincided with a sudden cultural awakening in Islam that resulted in a rapid absorption of the learning of the conquered neighbors of the Arabs. During the next century, the Abbasid caliphs al-Mansur, Haroun al-Raschid (of *Arabian Nights* fame), and al-Mamun made Baghdad into the cultural center of the world.

Among the scholars al-Mamun hired for the faculty of his "House of Wisdom" (*Bait al-hikma*) was the mathematician-astronomer Abu Mohammed ibn-Musa al-Khowarizmi (died ca. 850). His books on arithmetic and algebra played an important part in the subsequent development of mathematics in Western Europe. The translated title of the arithmetic book is *Concerning the Hindu Art of Reckoning,* and it is based on an Arabic translation of an earlier work from India, whence came our place-value system of numeration with nine digits. (The place-holder digit 0 was a later development and probably did not originate in India.) So thorough was al-Khowarizmi's explanation of arithmetical calculation that his book was the standard text in Europe (in Latin translation, of course) for centuries. This led to the false impression that our system of numeration is Arabic in origin, and we still speak of "Hindu-Arabic numerals."

Since European readers attributed the numeration to al-Khowarizmi, the system of calculation based on it was called by a corrupted version of his name, *algorismi,* later *algorism.* By association with arithmetic (which has a more legitimate derivation from the Greek word for number, *arithmos*), this got bent into *algorithm,* and today both algorism and algorithm appear in English dictionaries with a common meaning: the art of calculating with Hindu-Arabic numerals. But, as I have indicated, the latter word has also acquired a broader meaning, namely, any step-by-step procedure for solving a problem.

The Arabic title of al-Khowarizmi's most important book was *Al-jabr wa'l muqābalah,* and our word *algebra* is derived from the first word of the title. The book was, in fact, the first textbook on elementary algebra as it is studied today. It appears that "al-jabr" meant something like transposition of subtracted terms to the other side of an equation (negative coefficients were not allowed), and "muqābalah" apparently meant cancellation of like terms on opposite sides of the equation.[2]

The surname al-Khowarizmi actually refers to the birthplace of its owner, the town of Khowarizm. The town still exists, not much changed in appearance from the days when it was a center of Arab learning. Its modern name is Khiva, and it is located on the border between the Uzbek and Turkmen republics of the USSR. Because it still contains many of the structures of its glorious Arab past, the Soviet government has declared it a national monument and is in the process of restoring it.

References

Abercrombie; Boyer (1968); Gandz.

---

[2]A different view is expressed by Gandz (see Bibliography).

C H A P T E R 4

RABBITS, ROOTS, SINES, SUMS, AND TAXIS: SECOND STEPS IN PROGRAMMING

Each section of this chapter presents a problem that leads to a relatively simple programming problem. In some cases, variations of the programming problem are also suggested. It is not necessary that you do all the problems, but your instructor will probably want you to do at least two or three of them. You should at least read the others for the programming hints that they contain.

It is certainly not expected that you have already mastered the details of the programming language that has been introduced. As you try to write a program, refer to Chapter 1 as often as necessary. When a programming question arises for which you cannot find the answer in what you have learned so far, write down your best guess of what that program step should look like. If it happens not to be acceptable within the rules, the computer will let you know during your next laboratory session, and the correction can be made then.

4.1 TABULATION OF THE SINE FUNCTION

In the problem to be described in this section, we make use of a "built-in" function that is provided as part of your computer system, namely, the trigonometric *sine* function, whose name is SIN. If X is the name of an angle in radian measure, SIN(X) will give the value of the sine of that angle.

Problem 1 Make a table of the sine function for values of X from 0 to 6.5 radians (a little more than 360°) in steps of 0.5 radian.

Your flowchart and program require one loop in which there is an output statement to print values of X and SIN(X). You may want to label your output so that each line looks like

$$X = \underline{\hspace{1cm}} SIN X = \underline{\hspace{1cm}}$$

At some point in your mathematical career, you have probably seen and used a sine table. Now you can make your own (with a little help from your friend). During execution the computer is *not* "looking up" those values in some electronic table; it actually is computing them as fast as they appear on your screen or printer. Much later in the course (Chap. 37), you will learn at least one way this can be done (i.e., the kind of built-in programming that is abbreviated SIN).

Variations [O]

1. You may be more used to degree measure for angles than radian measure. Why not print each value of X in *both* measures? One radian is approximately 57.29578 degrees.

2. Another reasonable way to arrange the output is so that each line reads

$$\text{SIN} (\underline{\qquad}) = \underline{\qquad}$$

Can you do it?

3. Still another way to arrange output: Start your program with an output statement to print column headings:

$$X \qquad \text{SIN } X$$

Then enter the loop and have the repeated output step print only numerical values.

4. Why not make a more complete trig table with values of COS X and TAN X as well? COS(X) provides values of the cosine function. Your system may or may not recognize TAN(X). If it does not, use SIN(X)/COS(X) instead.

4.2 SOME SUMS

The introductory discussion of computers (Chap. 0) used as an example a hypothetical problem involving the computation of a very long sum. Sums are very important in our subsequent work, and this is a good time to learn the standard way to evaluate them. Suppose we want to add $a_1 + a_2 + a_3 + \cdots + a_m$, where the subscripts number the terms of the sum, and where each *term* a_k is to be computed by some formula involving k. We begin by assigning a variable name, say, S, which is used to accumulate the sum, and set $S = 0$. Then a loop indexed from 1 to m can be set up, with the body of the loop being an assignment statement that replaces S by S + (formula for the kth term), k being the index for the loop. The first time through the loop, the 0 in location S is replaced by a_1 (or $0 + a_1$). The next time, S is set to $a_1 + a_2$, then to $a_1 + a_2 + a_3$, and so on. When the loop has been completed, the value of S will be the sum of all the a's.

Problem 2   Write a program to evaluate *two* such sums, namely, the sum of the reciprocals of the first 1000 positive integers,

$$S = 1 + \frac{1}{2} + \frac{1}{3} + \frac{1}{4} + \cdots + \frac{1}{999} + \frac{1}{1000},$$

and the sum of the reciprocals of the *squares* of the first 1000 positive integers:

$$T = 1 + \frac{1}{4} + \frac{1}{9} + \frac{1}{16} + \cdots + \frac{1}{(999)^2} + \frac{1}{(1000)^2}.$$

Digression [R]   Later in your calculus course you learn that calculus enables you to do some remarkable things, such as "adding up" an *infinite* number of terms, (sometimes) getting a finite answer. A computer, even a very fast one working for a long time, can add up only a *finite* number of things. Nevertheless, when the sum of infinitely many terms actually is finite, after adding up some number of terms, you are already very close to the final answer, and "all the rest" of the terms can be ignored. Computers *can* do that kind of summation and hence can give very good approximations to sums of infinitely many terms, provided the sums are known to be finite, which is where the calculus comes in.

The sums you are to evaluate in this problem suggest consideration of the corresponding sums

$$1 + \frac{1}{2} + \frac{1}{3} + \frac{1}{4} + \cdots \tag{4.1}$$

and

$$1 + \frac{1}{2^2} + \frac{1}{3^2} + \frac{1}{4^2} + \cdots \tag{4.2}$$

where $\cdots$ in each case means "continuing on to infinity." Except for the accomplishments of Archimedes (287–212 B.C.) in summing an infinite geometrical progression, such sums were not even considered until the fourteenth century. One of the early accomplishments was the demonstration that the *harmonic series* (4.1) does *not* have a finite sum. This was done by Nicole Oresme (1323?–1382), a French mathematician and Bishop of Lisieux. On the other hand, (4.2) *does* have a finite sum. This fact was known in the late seventeenth century, but early attempts to *find* the sum failed. The first actual determination of the sum was done by the Swiss genius Léonhard Euler (1707–1783), who showed in 1736 that the sum (4.2) is $\pi^2/6$.

If the content of the preceding paragraph seems rather implausible to you, fine—just be patient! You learn later in your calculus course how Oresme showed that (4.1) does *not* have a finite sum, and also why it is easy to see that (4.2) does have one. Euler's result, that the sum is $\pi^2/6$, will not be found in the usual beginning calculus course, but if you want to see how he made the discovery (by some sloppy reasoning that involved pretending that infinite sums are just like finite ones), consult Boyer's *History,* pages 486–487. After a year or so of calculus, you will be prepared to study a real proof of Euler's formula, if you are interested. Consult the papers by Ayoub, Matsuoka, Stark, Giesy, and Papadimitriou listed in the Bibliography. Chapter 52 takes up the numerical determination of Euler's sum and other related sums.

4.3 RABBIT'S FRIENDS AND RELATIONS

Perhaps the earliest work in the field of mathematical ecology was the proposal and solution of a small problem in population dynamics in A.D. 1202 by Leonardo of Pisa, son of Bonacci, better known as Fibonacci (a contraction of *filius Bonacci*). Here is the problem: Starting with one pair of mature rabbits, how many rabbits will they produce by the end of a year if each pair gives birth to a new pair each month starting with the second month of its life and if no deaths occur?

The solution proceeds as follows. In the first month the starting pair produces one new pair, so there are two pairs at the end of the month. In the second month another new pair is produced—but the first offspring are not yet mature enough to produce—so there are three pairs at the end of the month. During the third month, the two mature pairs each produce a pair, for a total of 5. During the fourth month, there are three mature pairs—those born by the end of the second month—hence three new pairs for a total of 8. And so on.

What is happening here? In the nth month, the number of mature pairs (hence also the number of *new* pairs) is the total number alive at the end of the $(n-2)$th month. Hence in the sequence of monthly totals.

$$2, 3, 5, 8, 13, 21, 34, \ldots,$$

each number is the sum of the two preceding numbers. With this observation, you will have no difficulty extending the sequence to the twelfth entry to find Fibonacci's answer to the problem. (Do it!)

Obviously, we are working with an unrealistically simple mathematical model for rabbit populations, and, if that were all this computation is good for, we would probably not have bothered to mention Fibonacci. However, it turns out that the sequence of monthly totals, now known as the Fibonacci sequence, occurs in a variety of ways in nature and is useful in studies of such diverse phenomena as phyllotaxis (arrangement of leaves, buds, petals, and so on, in plants), proportions in musical composition, poetry, ancient mosaics, atomic physics, growth of cancers, and water pollution.[1]

To state our programming problem, we need some notation. It is conventional to push Fibonacci's problem back two months to account for the maturity of the original pair of rabbits. We write $F_1 = 1$, $F_2 = 1$, and thereafter

$$F_n = F_{n-1} + F_{n-2}. \tag{4.3}$$

(Hence $F_3 = 2$, $F_4 = 3$, $F_5 = 5$, and so on.) In addition to the Fibonacci numbers themselves, we will be interested in the ratios of successive terms:

$$R_n = \frac{F_n}{F_{n-1}},$$

for reasons that become clear later.

Problem 3   Write a program to tabulate N, F_N, and R_N for N ranging up to 25. Label each line of output as in Program 1.2 or Section 4.1.

HINT: To evaluate the "recursive" formula (4.3), you need three separate variables for Fibonacci numbers, say, $F1$, $F2$, $F3$. Each time through the loop you calculate $F3$ as $F1 + F2$ and R as $F3/F2$. Then before reaching the bottom of the loop, you must replace $F1$ by $F2$ and $F2$ by $F3$. Why?

4.4 BABYLONIAN SQUARE ROOTS

Reference was made in Section 4.1 to one of the "built-in" functions in your computer system, the sine function. There are a number of other very useful functions available to you,

---

[1]The study of the Fibonacci sequence and its relatives has become important enough to warrant publication of a scholarly journal called the *Fibonacci Quarterly*. You can read more about the matters mentioned here in various *FQ* articles, especially in Volume 1.

including one to calculate square roots.[2] Even though the square root function will be a great convenience in later programming, there is some satisfaction to be gained from learning how to compute square roots yourself. A method for doing this was known in ancient Mesopotamia[3] earlier than 1500 years B.C., so it can't be very difficult. In fact, this method is a lot easier than the one that used to be taught in American schools (perhaps still is in some schools), but it would require more arithmetic if carried out by hand. You probably have a calculator with a square root key, but if you had only a "four-function" calculator, you could still compute square roots by the method presented here.

Suppose we want to find $x = \sqrt{a}$, where a is some positive number. We begin with a first *guess,* say, x_1. (Any guess, no matter how crude, will do, as long as it is positive. However, if arithmetic is being done by hand rather than by using a computer, a "good" guess will shorten the computation considerably.) Now compare x_1 with a/x_1. The product of these two numbers is a, so if x_1 is smaller than $\sqrt{a}$, a/x_1 must be bigger than $\sqrt{a}$, and vice versa. (Why?) That is, $\sqrt{a}$ must lie *between* x_1 and a/x_1. It follows that the *average* of these numbers,

$$x_2 = \frac{(x_1 + a/x_1)}{2},$$

must be a better approximation than at least one of them. (Why?) Now $\sqrt{a}$ must also lie between x_2 and a/x_2 (for the same reason as before). Thus we average these to get a still better approximation:

$$x_3 = \frac{(x_2 + a/x_2)}{2}.$$

The repetitive nature of the task is now clear. As often as we like, we can compute a "new x" by averaging the "old x" and $a/$"old x." At each step, $\sqrt{a}$ lies between x and a/x, and the distance between these two gets smaller at each step.

Here is an example of such a computation, in effect a translation from an old Babylonian clay tablet. (The original used cuneiform numerals and sexagesimal fractions.) We start with $a = 2$ and a guess of $x_1 = 1.5$ as our first approximation to $\sqrt{2}$.

$$x_1 = 1.5, \qquad\qquad\qquad\qquad\qquad 2/x_1 = 1.333333$$
$$x_2 = (1.5 + 1.333333)/2 = 1.416667, \qquad 2/x_2 = 1.411765$$
$$x_3 = (1.416667 + 1.411765)/2 = 1.414216 \qquad 2/x_3 = 1.414211$$

(All calculations were rounded to six decimal places.) Because x_3 and $2/x_3$ agree to five decimal places and $\sqrt{2}$ lies between them, we must have $\sqrt{2} = 1.41421$, correct to the fifth decimal place.

Besides showing that the method is quick (given a good starting value), this computation also illustrates how to decide when to stop the repetitive step (loop). For example, for five correct decimal places in the answer, we can stop when

$$\left| x - \frac{a}{x} \right| < 0.00001.$$

---

[2]The name of that function is SQR in BASIC.

[3]The Mesopotamian civilization is sometimes called, not quite correctly, "Babylonian." See Boyer (1968), Chapter 3.

(The absolute value of the difference is necessary because we cannot tell yet whether x or a/x will be larger at any given step, and *every* negative number is smaller than 0.00001.)

Problem 4   Write a "square root" program to compute $\sqrt{A}$ to five decimal places, where A is an input variable. The initial guess may also be an input item, or you may write in a "guess" that is independent of A, say, $X = 1$. Have your program print *all* the successive values of X.

Here is a small hint for writing the program. The end test suggested earlier is similar to the one in Program 1.2. Thus the main loop may be constructed with an upper index large enough that it should never be reached, with an interruption of the loop when x and a/x are close enough to each other. To translate the end test into the conditional part of an IF statement, you should know about the *absolute value* function ABS: If E is any expression, the value of ABS(E) is the absolute value of the value of E.

4.5 HARDY'S TAXI

The following story is sometimes cited as an illustration of the remarkable arithmetic ability of the self-taught Indian genius Srinivasa Ramanujan (1887–1920), although it may really say more about his memory for numerical facts.

While hospitalized in England, Ramanujan was visited by his friend, the British mathematician G. H. Hardy. Hardy happened to mention that he had arrived at the hospital in a taxi with the "dull" number 1729. Ramanujan immediately responded that the number is not at all dull because it is the smallest positive integer that can be written in two different ways as the sum of two cubes.

Problem 5   Write a program to verify Ramanujan's claim—i.e., $1729 = m^3 + n^3$ in two different ways (not just interchanging m and n) and no smaller number has this property.

Here is a suggestion for how to proceed. You can check by hand that 13^3 is more than 2000, so neither m nor n can be that large. Then make a table of values of m, n, and $m^3 + n^3$ for m and n, ranging from 1 through 12, with n always $\geqslant m$. (Why do we want this last restriction?) If the result is correct, 1729 should appear twice in the third column, and no smaller number should appear more than once.

Because both m and n are varying here, this will provide your first introduction to "nested loops," i.e., one loop contained in another. The main portion of your flowchart should look something like Figure 4.1.

NOTE: Effective use of this program requires a printer or printing terminal, since the tabulation is too long to fit on a video screen, and you need to be able to see it all at once. More sophisticated analysis and programming (see the next section) makes a printer unnecessary.

4.6 WHAT TO DO IF YOU ARE BORED [0]

In this section we present two problems that are variations of the problems presented in Sections 4.3 and 4.5 but which require more sophisticated analysis and programming. The second problem requires knowledge of programming with "arrays" or "subscripted variables," which will not be discussed or otherwise used in this course. Such knowledge may also simplify the programming of the first problem, but it is not required.

Figure 4.1

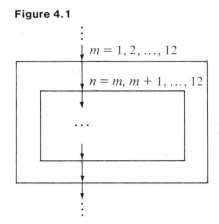

Problem 6   Among the many flaws in the Fibonacci problem (Sect. 4.3) as a model for the population dynamics of rabbits is the rather obvious fact that no provision is made for the death of any of the rabbits. Suppose we make an additional assumption that rabbits die when they are exactly a year old. Write a program that will produce a month-by-month tabulation of immature, mature, and dying rabbits, and total number of pairs of rabbits for a two-year period starting with one pair of immature rabbits.

NOTE: Each pair produces a new pair in its twelfth month, just before it dies.

The "dying-rabbit problem" was posed by Brother U. Alfred in the first issue of *Fibonacci Quarterly,* and he subsequently proposed a solution in the fourth issue of Volume 1. (That article also contains a two-year table like the one asked for in this problem.) An error in his solution was later pointed out by J. H. E. Cohn, who gave a correct solution. More recently, V. E. Hoggatt, Jr., and D. A. Lind have given a complete solution of the dying-rabbit problem for arbitrary breeding patterns and death times.

The problem is still very artificial from the point of view of population dynamics. With the recent upsurge of interest in ecology, this has become a lively field of inquiry. However, to understand the dynamics of birth-death processes, predator-prey situations, and so on, a substantial background in calculus and differential equations is required. That may be one reason you are taking this course (even if you have not thought about it that way previously). These subjects are taken up in Chapters 48 through 51.

Problem 7   Our statement of Hardy's taxi problem in Section 4.5 called for producing a lengthy table of sums of cubes that would then be checked by hand to see if 1729 appears twice and to determine whether any smaller number does. Write a program that will do the whole job—i.e., will determine whether Ramanujan's claim was correct.

References

General: Boyer (1968). Section 4.2: Ayoub; Giesy; Matsuoka; Papadimitriou; Stark. Section 4.3: Coxeter; Dean; Gardner (1959, 1969); Horadam; von Baravelle (1948). Section 4.4: Jones (1949); Kovach; Neugebauer (Chap. 2). Section 4.5: Gardner (1973b); Newman. Section 4.6: Alfred (1963a, 1963b); Cohn; Hoggatt and Lind.

C H A P T E R 5

RUNNING YOUR OWN PROGRAMS [L]

Once you have prepared one or more programs in response to the problems in Chapter 4, you will want to try them out. This chapter provides some useful information to have in hand during your next session. We recommend that you read through Section 5.0 beforehand, as well as the sections that pertain to the specific programs you will be trying.

5.0 EXTERMINATION OF BUGS

By now you will have discovered that analysis of a problem, preparation of an algorithm for the solution, and preparation of a program for the algorithm all require that considerable attention be paid to *details,* even if the problem to be solved is a rather modest one. With so many details to worry about, occasional errors are inevitable, especially as the problems become more complex. All sorts of errors—whether the result of faulty logic in the program design, misuse of the programming language, or failure to type what was intended—are called *bugs,* and the process of locating and exterminating bugs is called *debugging.* (Several manufacturers of computer systems have used the acronym DDT for programmed debugging aids—e.g. to stand for Dynamic Debugging Technique.)

Even though the problems in Chapter 4 call for fairly short and simple programs, it is possible that you will need to do some debugging in this session. There are no hard-and-fast rules for doing this, but we begin with some helpful hints for debugging, just in case you encounter difficulty. The remainder of the chapter consists of comments on what to look for in running each of the programs from Chapter 4. You may ignore comments pertaining to any programs you have not written.

The simplest errors to find and correct are typing mistakes. The following paragraphs review briefly what can be done about them. (See Sect. 2.0 and the corresponding information about your system provided by your instructor.)

1. When one or more incorrect characters are typed and recognized as such before the line is entered, you may cancel them or backspace over them and retype the rest of the line (Sect. 2.0, item 11).

2. When an error is recognized before the line is entered and you don't want to cancel or backspace—e.g., if the error is near the beginning of a long line—the entire line can be wiped out (Sect. 2.0, item 12) and then retyped.

3. If an incorrect line is entered and is accepted as a valid statement, typing the correct line with the same line number wipes out the incorrect line.

NOTE: Some systems do not check lines for valid syntax as they are entered, but they do after a RUN command has been given.

4. If your system checks for syntax errors as lines are entered, an incorrect line may provoke a response from the computer in the form of an error message. (Even if this checking is

done when you RUN, this remark will still apply at that time.) If you recognize what is in error, simply type in the correct line with the same line number.

There is, of course, the possibility that you will not know the correct statement for doing what you wanted to do. In that case there are several possible sources for the information you need: Chapter 1 of this text, a user's manual (if one is available), or a friendly neighbor in the computer room. (All computer users are friendly to each other because they all have a common adversary.) If these sources fail you, you should face up to the possibility that you have not understood something important, in which case you consult your instructor.

The difficult bugs to locate are those that result from faulty logic in the design of the program. These show up only when you try to RUN a program that consists of valid statements, as judged by the processor, and the program fails to produce right answers. (This may mean that it produces no answers, wrong answers, or an error message.)

An ounce of prevention is sometimes worth an hour of cure. As we have emphasized previously, before executing an apparently valid program, you should always LIST it, and then study the listing to make sure it looks like what you intended. Also check to make sure each variable in the program is assigned a value somewhere earlier in the program than each place where it is expected to have a value. Have your flowchart at hand, and check again that the program accurately reflects the logic of the chart. Finally, check that each possible route through the program contains some output step.

When you are ready to run your program, it is important that you be able to tell what constitutes a *right* answer. What this means depends on the problem at hand, and we provide additional comments on this in connection with specific problems. In general, if your program has input variables, you should check it first with input values for which you know the right answers or can easily check by hand computation. If your program has many branch points (decision steps) in it, you may need several test sets of input data to check all the possibilities. (We have not considered any such problems yet, however.)

Now suppose that, in spite of your best preventative efforts, you fail to get right answers. There is one possibility that may call for emergency action, namely, no response at all to the RUN command. Assuming you have made sure that it is impossible to go all the way through your list of instructions without reaching either an output or a STOP statement, if the computer sits quietly for more than a few seconds, this may mean trouble. If possible, get the computer's attention (Sect. 2.0, item 16; BREAK, RESET, ATTN, or STOP key). If you have inadvertently tied up the computer in an infinite loop, execution will cease, and the system will identify the line on which it was last working. If you are using a time-sharing terminal, it is possible that your connection to the computer was somehow broken, in which case you will get no response to your action, which means you should log on again. Some terminals have a light to indicate when you have a connection, whereas others do not have a light.

A second possible alternative to right answers is an error message. The message will indicate what went wrong, such as the failure of a variable to have a value when needed, an attempted division by zero, demand for the square root of a negative number, generation of too large a number ("overflow"), and so on. It may also identify the line number of the offending statement. The question, "How could that have happened at that point in the program?" sometimes has an obvious answer, in which case you have located the bug.

The third, and most frightening, alternative to right answers is *wrong* answers. Again, it is possible that the wrong answer itself makes the location of the bug obvious, but more often it only creates a mystery. Whether your program terminates with an interruption, an error message, or a wrong answer, if you do not quickly spot the bug, there are several tools you can use to solve the mystery.

1. If your system permits direct inquiries, you can ask for the last known values of your variables, loop indices, and so on.

2. You can selectively "trace" your program. In a suspect portion of the program, insert an output statement after every step to get the values of all the relevant variables. Then execute again. Or use the "trace mode," if your computer has one.

3. You can vary the input data to see if a different case leads to the same type of error.

4. If the mystery remains, you can retire to a quiet room to contemplate your program LIST, your flowchart, your various traces and direct inquiries, and your wounded pride. At this point you may need help—see your instructor.

It is not really expected that debugging programs from Chapter 4 will require any of the more extreme measures just indicated. However, as the logic of your programs increases in complexity, the opportunities for mistakes also increase, and you may need to refer to this section later.

5.1 PRODUCING A SINE TABLE

This section is your "lab manual" to accompany attempted execution of the program called for in Section 4.1. After you have turned on your computer, the next step is to type in your prepared program. Before trying to run it, obtain a listing of what the computer thinks the program is, as noted in the previous section. After corrections, if any, give the appropriate command to the computer to execute your program. This program has no input variables, and if everything works right, your output should be a 14-line table of pairs of numbers. Is that what you have?

Check to see that your answers make sense. The x-values (representing angles in radian measure) should read 0, 0.50, and so on up to 6.50. The corresponding values of sin x should all be between -1 and 1. Do they start from 0, increase to about 1, and then start decreasing when x is near $\pi/2$ (the radian measure of 90°, about 1.57)?

If your program incorporates any of the variations suggested in Section 4.1, check to see that what you intended was actually carried out. In particular, if you are using column headings, you may not be satisfied with the centering of them. You can adjust the corresponding output statement by inserting or deleting an appropriate number of blanks and then running the program again.

If you have a basic program that works right, you may want to try some of the variations from Section 4.1 now by modifying the program. Feel free.

Finally, we have an additional variation to suggest to users of BASIC. As it stands, your program may have a loop governed by an index that takes values from 1 to 14 (or perhaps 0 to 13). However, a feature of the loop-building FOR statement makes it possible to use X itself as the index, with the step size from one value of X to the next equal to 0.5. The appropriate statement is

$$\text{FOR X} = 0 \text{ TO } 6.5 \text{ STEP } .5$$

with the loop ending with

NEXT X

When STEP is omitted in FOR statements, as we have done previously, the increment is assumed to be 1, but, as you can see, you are not limited to that choice. Using the loop statement just described, you may now reduce the body of your loop to a single statement, namely, an output statement to print X and SIN(X). Give it a try.

5.2 SUMS OF RECIPROCALS

A program for Section 4.2 requires only a few lines. It may have occurred to you, as both sums to be computed are of the same length, to write just one loop, indexed from 1 to 1000, in which S would be increased by $1/K$ and T by $1/(K*K)$.

NOTE: $K*K$ is mathematically equivalent to K squared.

When you run your program, you should get two answers. The answer for T should be easy to check because (according to Euler) it should not be too far from $\pi^2/6$. Try entering and running a one-line program with an output statement to print the constant $(3.14159 \uparrow 2)/6$ (or use a direct inquiry or your personal calculator). Is it close to T?

The answer for S may be a little surprising, because (according to Oresme) it should be the sum of the first thousand terms of a sum that is "eventually infinite." S is, in fact, not very large. Yet adding additional terms, all of which are less than 0.001, will eventually produce a sum as large as you can imagine. On the other hand, adding additional terms of the form $1/k^2$ to T (even infinitely many additional terms) does not change it very much. The mystery in all this will be dispelled later in your study of calculus.

5.3 FIBONACCI NUMBERS

A program written in response to Section 4.3 should produce a three-column table of up to 25 lines. (It might be shorter if you start printing at $N = 2$ or 3, say.) Your table may not fit on your display if you are using a 24-line video screen, but it won't hurt anything if you lose a few lines at the top. It also won't hurt if you want to shorten your loop by a few steps. Here are a few sample values to check:

$$F_{12} = 144 \qquad F_{19} = 4181 \qquad F_{25} = 75025.$$

(If you did not get these values or if they turned up with the wrong values of N, debugging is in order. See Sect. 5.0.)

Assuming that your Fibonacci sequence (the F-column) is in order, let's look at the R-column, which should contain ratios of consecutive Fibonacci numbers. If correctly computed, this sequence of numbers should have a striking property: As N increases, the ratios all begin to look alike! In fact, they should get closer and closer to a particular number that starts out 1.618. . . . (If your R-column does not look like this, you have another bug to find.) Let's determine why this happens. The basic property of Fibonacci numbers is the relation

$$F_n = F_{n-1} + F_{n-2}.$$

If we divide through by F_{n-1}, we get

$$\frac{F_n}{F_{n-1}} = 1 + \frac{F_{n-2}}{F_{n-1}},$$

or

$$R_n = 1 + \frac{1}{R_{n-1}}.$$

If there really is a number R to which all the R_n's get close as n gets larger, this equation suggests that R must satisfy

$$R = 1 + \frac{1}{R}.$$

(Why?) If we multiply through by R and rearrange terms, the equation becomes

$$R^2 - R - 1 = 0,$$

which has the two solutions $(1 + \sqrt{5})/2$ and $(1 - \sqrt{5})/2$. (Why?) R cannot possibly be negative. (Why?) So *if* there is such a number R, it must be $(1 + \sqrt{5})/2$.[1] To check whether the R_n's really are close to that number, you can enter and run a one-line program to print the value of that constant (or use a direct inquiry or your calculator). In BASIC the output line is

$$\text{PRINT } (1 + \text{SQR}(5))/2$$

5.4 SQUARE ROOTS

Your Babylonian square root program should accept any positive number as an input value for A and produce a sequence of answers, the last of which is $\sqrt{A}$. To check it out, let's start with a problem for which we know the answer: $A = 4$. (If your initial guess is also an input variable, *don't* take it to be 2. Use 1 instead.) Any perfect square will do for a test case. You can also try to duplicate the computation of $\sqrt{2}$ given as an example in Chapter 4. Does your program work? If not, now is the time to learn about debugging (see Sect. 5.0). If it does work, you can execute it as often as you like with other choices of A. You may want to experiment to see what happens when the initial "guess" is not very close to the right answer.

5.5 SUMS OF CUBES

If you have carefully followed the instructions in Section 4.5, you have a program with a "nested" pair of loops, the body of which should be an output statement to print m, n, and $m^3 + n^3$. Execution of this program should produce a table 78 lines long.

---

[1]This number is the ratio involved in the so-called Golden Section, the division of a line segment into two parts so that the ratio of the whole segment to the larger part equals the ratio of the larger to the smaller part. This proportion has a fascinating history, which can be traced back to Greek geometry of the sixth century B.C., the time of Pythagoras (250 years before Euclid). For an excellent exposition, including the connection with Fibonacci numbers, see Coxeter.

Recall what it is that the table is supposed to show: The smallest number that appears twice in the third column (possibly the only one that appears twice) should be 1729. The corresponding numbers in the first two columns show the two different ways 1729 can be written as the sum of cubes. Was Ramanujan right?

C H A P T E R 6

FUNCTIONS AND GRAPHS [L]

6.1 FUNCTION DEFINITION

In this chapter you find our first application of the computer to the calculus course, one that will be useful to you from time to time throughout the course. By now you have seen a variety of (mathematical) functions, in both algebraic and graphical form. We have also made reference in earlier chapters of this book to built-in functions in your computer system, such as sine, square root, and absolute value. Each of these really is a *function* in the mathematical sense: To each value of the independent variable (the thing in parentheses following the function name), there corresponds exactly one function value. (If necessary, translate this statement into the terminology used in your calculus book for defining function.) However, you will often find it necessary to use functions other than the ones provided by your system. You may define your own functions in programs by means of function definition statements, which we describe now.

Suppose we want to make frequent reference in a program to the polynomial function

$$F(x) = x^4 - 2x^3 + x^2 + 6x - 7.$$

This may be done by inserting the following statement in the program:

100 DEF FNF(X) = X↑4 − 2∗X↑3 + X∗X + 6∗X − 7

All function names in BASIC have exactly three characters. For user-defined functions, the first two characters must be FN, and the third must be a letter, so there are 26 possible defined function names:

FNA, FNB, . . . , FNZ.

On some computer systems, DEF statements may appear anywhere in a program and do not necessarily have to come before the first use of the functions they define. On others they must precede use of the functions. It is good practice (and valid in virtually all BASICs) to put DEFs near the start of the program in a portion of the code that is executed only once (i.e., before all loops). That assures DEFs will: (1) be easy to find and modify; (2) precede use of the functions, in case that is required; and (3) be "executed" only once, as some systems require.

Once a function FNF has been defined, that name can be used for the function throughout the program in exactly the same way SIN is used for the sine function or ABS for the absolute value function. For example, if A and B are variables that have been assigned values, and you want to compute the average of $F(A)$ and $F(B)$, you could write

$$500 \text{ LET } G = (\text{FNF}(A) + \text{FNF}(B))/2$$

Any arithmetic expression can be used as an argument for such a function, and one such function can be used in defining another or in forming arithmetic expressions.

The variable used in defining a function (e.g., X in the preceding line-100 example) is a "dummy" variable that has no connection with use of the same symbol elsewhere in the program. Exactly the same function would be defined in line 100 if X were replaced by T, say, throughout the line. (If you understand the mathematical definition of function from your calculus course, that last assertion will be obvious to you.)

BASICs differ on how many arguments a function can have, i.e., on whether you can DEF a function of two or more variables, say, FNF(X,Y). We do not have any need for such functions in this course anyway. However, most BASICs allow variables other than the argument(s) to appear in the expression on the right; whenever such a function is called, the current values of those variables are used. For example, if you wanted to define a general quadratic function, you could write

$$110 \text{ DEF FNQ}(X) = A*X*X + B*X + C$$

and this would define the entire infinite family of such functions. Each time FNQ was used, you would have to be sure that A, B, and C had the values you wanted for coefficients, and as before, you could substitute any expression for X.

As with variable names in general, the choice of function names is entirely up to you, subject to the limitations as previously noted, but it is good practice to use names that are meaningful to you and easily recognized later. For example, if you have defined a function FNF and also want to use its *derivative* in your program, you might want to name the derivative function FND.

6.2 TABULATION OF FUNCTIONS

One of the ways a computer can help us study functions is by tabulating them—i.e., producing a list of function values at x-values ranging over an interval (a, b) at some specified spacing. We saw an example of this in Sections 4.1 and 5.1 that involved tabulation of the sine function. Of course, a finite table can never be a complete picture of a function that has infinitely many numbers in its domain, but if the function is reasonably well behaved (as are the functions usually studied in a calculus course), a table can show us where the function is increasing or decreasing (as we go from left to right along the x-axis), where it takes values near 0, where it is positive rather than negative, and so on.

Figure 6.1 shows a general-purpose function-tabulating program and its flowchart. Once you complete the definition of the function, execution of the program will produce a table of

Figure 6.1

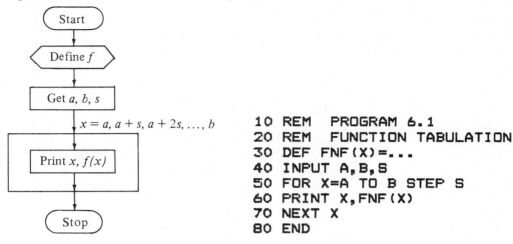

```
10 REM   PROGRAM 6.1
20 REM   FUNCTION TABULATION
30 DEF FNF(X)=...
40 INPUT A,B,S
50 FOR X=A TO B STEP S
60 PRINT X,FNF(X)
70 NEXT X
80 END
```

values of x and $f(x)$, where x goes from a to b with a spacing of s. (Note the use of a different shaped box in the flowchart to indicate a definition.)

As preparation for the exercises that follow, you must have the appropriate version of Program 6.1, except for line 30, either punched on cards or typed in at your terminal. For each exercise you must insert an appropriate line 30 (function definition statement) before running the program. Batch-process users also need different cards for each run.

Exercises

Tabulate each of the following functions on the indicated intervals and with the indicated spacing. Examine the tabulation for each to see what it tells you about the function. Plot your results on graph paper to obtain a graph of each function.

NOTE: Whether you are using a video display or a printer for output makes a difference in these exercises. Printed output enables you to take your tabulations away from the computer to draw your graphs (and for future reference). However, even with a video display, you can get enough information to draw a reasonable graph in the following way: Copy down by hand every other X, Y pair; note that you only need the first two or three digits of each Y. The first five exercises produce tabulations that fit on a screen. Exercises 6 and 7 have to be done in segments or with a larger step size.

1. $f(x) = (2 - x)/(1 + x)$ on $[0, 4]$, $s = 0.25$.

2. $f(x) = x^3 - 3x + 3$ on $[-2, 2]$, $s = 0.25$.

3. $f(x) = -26/(x + 2) - (x/6)^3 + x + 13$ on $[0, 20]$, $s = 1$.

4. $f(x) = x/(x - 1)$ on $[-2, 4]$, $s = 0.5$.

5. $f(x) = x/(x - 1)$ on $[-2, 4]$, $s = 0.45$.

6. $f(x) = x^4 + 2x^3 - 6x^2 + x - 7$ on $[-4, 4]$,　$s = 0.25$.

7. $f(x) = (x^4 + x - 1)/(x^3 + 2x^2 - 1)$ on $[-4, 3]$,　$s = 0.2$. Also tabulate the numerator and denominator separately. Determine smaller intervals that seem to be of greater interest, and obtain finer tabulations (say, with $s = 0.02$) on those smaller intervals.

8. $f(x) = |x|/(x^2 + 1)$. Choose your own interval(s) and step size(s).

From time to time later in the course you may want to get some preliminary information about the behavior of some function, such as where does it change from positive to negative values? Any time you are at the terminal, it will take only a few minutes to type in Program 6.1 (or reuse your card deck) and experiment with the function of interest. You might want to start with a very coarse tabulation over a large interval, and then compute finer tabulations over smaller intervals that look interesting.

6.3 A PICTURE IS WORTH A THOUSAND NUMBERS

There is an old saying that "A picture is worth a thousand words." A saying gets to be an *old* saying by being (largely) *true*. And it is also true that a picture, such as the graph of a function, can be worth far more than a thousand numbers. Some programs to draw pictures for you have been prepared in connection with this text, and we now describe how they work.

There are essentially two distinct ways a computer can draw a picture: by "characters" and by "dots." Character graphics are available on virtually all computers. In this mode, pictures are built out of the characters that are available on the keyboard; some computers have special graphics characters for this purpose, others use just the typewriter characters. The resolution of character graphics is severely limited by the physical limitations on where characters can be placed on the output device (e.g., 24 lines of 40 characters on a video display). In particular, the typical microcomputer video display is simply inadequate for drawing a reasonable graph of a function in character mode. On the other hand, quite good graphs can be drawn on printers in character mode if the line width is, say, 80 or more characters.

Dot graphics are based on the obvious fact that characters (and other images) on a video screen (and also on "dot matrix" printers) are made up of many little dots (called "pixels" in video parlance). If the computer has the capability of controlling the dots individually, pictures can be drawn on the screen or printer in much higher resolution than is possible with character graphics. For example, the typical microcomputer screen of 24 40-character lines may have about 200 lines of about 300 dots each.

Many computer systems have dedicated graphics devices with much higher resolution— e.g., graphics terminals or pen plotters. If you have access to such a device, you don't need the programs to be described here. As hardware prices continue to drop, these devices will become relatively common, but they are not common in educational computing environments yet, so we will say no more about them. (Well, just a little more. Plotters are based on a third mode, namely, drawing continuous lines that are not made up of individual dots, so resolution is limited only by the pen width and the accuracy of the motors that control pen position. They can also draw characters, but much more slowly than dot matrix printers. The educational computing configuration of the not-too-distant future will consist of a micro-computer with video monitor, a dot matrix printer, *and* a pen plotter, each output device being used for what it does best, and at a total cost not much higher than that of a terminal to a time-sharing system only a few years ago.)

Because of the limitations on commonly available output devices, we provide programs for graphing in two modes: character graphics on printers with at least 80-character line width (whether driven by micros, time-sharing, or batch systems) and video dot graphics for the most popular microcomputers. The programs are called GRAPH.LR and GRAPH.HR for "low resolution" and "high resolution," respectively. Elsewhere in this book we refer to both programs as GRAPH, and you may use whichever is appropriate for your system.

GRAPH.LR[1] is listed in "standard" BASIC in the Instructor's Manual. It may be typed in and saved in a disk file or on cards for time-sharing or batch systems. It is also provided on diskette for supported microcomputers with popular printer configurations. Any particular configuration may require minor modifications in the program. Here is a brief description of what the program does:

The physical limitations of printers (without dot graphics capabilities) dictate that we should take our y-axis horizontal on the page and our x-axis vertical. The process of plotting a graph from left to right, in the usual orientation, is then accomplished by letting each line of type correspond to an x-value and by plotting a point an appropriate distance along the line, as determined by the corresponding y-value. When the finished product is rotated 90° to the left, it will look like a graph in the ordinary sense. You are expected to enter your function (after LOADing the program) by a DEF FNF statement on line 10. When you RUN the program, you are asked for an x-interval $[A, B]$ and a step size S for x. The program then computes the smallest and largest y-values to be plotted (C and D, respectively). The interval $[C, D]$ will be subdivided into 50 parts (corresponding to 50 distinct character locations), and each plotted point will be represented by an appropriate mark in the nearest character space to the actual point. A labeled y-axis and a list of all the (x, y) pairs used will also be printed.

NOTE: Those with printers capable of 120 or more characters per line may wish to change the 50 subdivisions to 100 for better resolution.

GRAPH.HR[2] is provided on diskette for supported microcomputers. It draws graphs as discrete plotted points on the high-resolution screen with the usual x, y orientation. It too expects entry of the function by DEF FNF on line 10 and entry of endpoints A and B and step size S in response to INPUT. It scales both the x- and y-axes automatically according to the resolution of the particular micro and presents a graph in the usual orientation on the high-resolution screen. The space bar on the keyboard may be used to toggle back and forth between the picture and a text screen that describes the scaling. (Coordinates of the individual points are not printed, but you may use Program 6.1 to get them.)

---

[1]This program has been adapted, with permission, from the FORTRAN subroutine PLOT found in D.E. McLaughlin, *Numerical Solution of Ordinary Differential Equations,* Iowa Curriculum Development Project, University of Iowa, Iowa City, 1973, p. 145.

[2]This program is adapted from the graphical component of MATHPROGRAM, Copyright 1982, 1983, by R.C. Barr, T.M. Gallie, M.J. Hodel, R.E. Hodel, F.J. Murray, D.A. Smith, and D.A. Smith, II, published by CONDUIT.

Exercises

Use GRAPH to plot the same functions you tabulated with Program 6.1. The same values for A, B, and S may be used in each case.

6.4 SOME PICTURES ARE WORTHLESS [0]

This exercise illustrates graphing a function may actually lead to *confusion* unless the spacing is chosen carefully.[3]

1. Enter the following function with an appropriate function definition statement in the GRAPH program:

$$F(x) = x \sin \frac{\pi(1 + 20x)}{2}.$$

(Take π = 3.14159 throughout this problem.)

2. Graph F on the interval [0, 1], taking S = 0.2. What does the graph look like?

3. Perhaps the spacing is not small enough. Run the program again, taking S = 0.1. Now what does the graph look like?

4. Let's try again. This time take S = 0.05. Do you feel that you now have a better idea of what the graph *really* looks like?

5. One last try: Halve the spacing again by setting S = 0.025. Does that help? All the previous graphs are included in this one. Can you see what was causing the confusion?

---

[3]This example is adapted, with permission, from W.S. Dorn, G.G. Bitter, and D.L. Hector, *Computer Applications for Calculus*, Prindle, Weber, and Schmidt, Boston, 1972, pp. 12–20.

C H A P T E R 7

INFINITE LISTS OF NUMBERS

7.0 WHY WE NEED THIS CHAPTER

The word "sequence" has been used informally in several places in previous chapters of this text. At this point we need to take a closer mathematical look at the concept represented by this word, because sequences turn out to be a very important part of almost everything we do from now on. The study of sequences and their limits comes rather late in most standard calculus courses (toward the end of the first year or beginning of the second) which is unfortunate from our point of view. This arrangement is not, however, the result of its being a difficult topic or one that requires elaborate preparation—quite the contrary, as we will soon see. The real reason is that the traditional structure of courses using calculus (physics, engineering, chemistry, economics, etc.) calls for students to learn the important concepts of *derivative* and *integral* as early as possible and to spend a lot of time learning to use them in nonnumerical computations. Each of these concepts is a certain kind of *limit,* which is the fundamental concept of all calculus. Besides the fact that limits of sequences are fundamental for intelligent use of a computer, they are also the simplest kind of limit one can study mathematically. As undergraduate courses in mathematics and its applications become more computer-oriented, the study of limits of sequences will gradually be pushed forward to the beginning of the calculus course. If this has not yet happened in your course, this chapter establishes a few basic ideas that reappear in later chapters.[1]

7.1 SEQUENCES

For most purposes, the following definition is adequate for describing what is meant by sequence.

Definition   A *sequence* is a function whose domain is the set of positive integers.

If S is the name of such a function, then we can "list" its values:

$$S(1), S(2), S(3), \ldots .$$

When dealing with this very special kind of function, it is conventional to change the notation so that elements of the domain are written as subscripts rather than in parentheses:

$$S_1, S_2, S_3, \ldots .$$

Thus we may think of a sequence as an infinite list of numbers: a first one (S_1), then a second one (S_2), then a third one (S_3), and so on.

Sometimes it is convenient to change "positive" in the preceding definition to "nonnega-

---

[1]See Thomas for a complete but elementary development of limit concepts based on limits of sequences.

tive," so that a sequence may be numbered

$$S_0, S_1, S_2, \ldots.$$

This is still just an infinite list of numbers, but the first one is now called S_0, the second S_1, and so on. More rarely, we may want to allow *subsets* of the nonnegative integers as domains when it is more convenient to label our list as, e.g.,

$$S_0, S_2, S_4, S_6, \ldots,$$

or

$$S_1, S_3, S_5, S_7, \ldots,$$

or

$$S_1, S_2, S_4, S_8, \ldots.$$

The concept of an infinite list of numbers remains the same, however, regardless of the choice of numbering for the entries in the list. If it is understood which integers n are in the domain of a given sequence S, the sequence is sometimes specified by giving an explicit formula for S_n as a function of n. For example, $S_n = 1/n$ represents the sequence

$$1, \frac{1}{2}, \frac{1}{3}, \frac{1}{4}, \ldots.$$

Note that the numbering could not begin with zero in this case. (Why?)

Here are several examples of sequences:

Example 1 $S_n = n$: $1, 2, 3, 4, 5, \ldots$

Example 2 $S_n = 1 + \dfrac{(-1)^n}{2^n}$: $\dfrac{1}{2}, \dfrac{5}{4}, \dfrac{7}{8}, \dfrac{17}{16}, \dfrac{31}{32}, \ldots$

Example 3 $S_n = \dfrac{3}{10} + \dfrac{3}{100} + \cdots + \dfrac{3}{10^n}$: $0.3, 0.33, 0.333, 0.3333, \ldots$

Example 4 $S_n =$ the nth decimal approximation to π (there is no explicit formula for this):

$$3.1, 3.14, 3.141, 3.1415, 3.14159, \ldots$$

Example 5 $S_n = \pi$ *rounded* to n decimal places:

$$3.1, 3.14, 3.142, 3.1416, 3.14159, \ldots$$

The following examples first appeared in Chapter 4:

Example 6 $S_n = 1 + \dfrac{1}{2} + \cdots + \dfrac{1}{n}$: $1, \dfrac{3}{2}, \dfrac{11}{6}, \dfrac{25}{12}, \dfrac{137}{60}, \ldots$

(See Sect. 4.2. This sequence may also be defined "recursively" by $S_1 = 1$ and thereafter $S_n = S_{n-1} + 1/n$.)

Example 7 $S_n = 1 + \dfrac{1}{4} + \cdots + \dfrac{1}{n^2}$: $1, \dfrac{5}{4}, \dfrac{49}{36}, \dfrac{205}{144}, \ldots$

(Again, see Sect. 4.2. How would you define this sequence recursively?)

Example 8 $F_1 = 1, F_2 = 1$;

thereafter,

$$F_n = F_{n-1} + F_{n-2}:$$

$$1, 1, 2, 3, 5, 8, 13, 21, \ldots$$

Example 8 is the Fibonacci sequence of Section 4.3. At present, we know only a recursive definition for this sequence. There is actually an explicit formula for the nth term, which looks most implausible in view of the fact that the terms are all integers:

$$F_n = \frac{1}{\sqrt{5}} \left[\left(\frac{1 + \sqrt{5}}{2} \right)^n - \left(\frac{1 - \sqrt{5}}{2} \right)^n \right]$$

Example 9 $R_n = F_n / F_{n-1}$ for $n \geq 2$: $1, 2, \dfrac{3}{2}, \dfrac{5}{3}, \dfrac{8}{5}, \dfrac{13}{8}, \ldots$

Example 9 is the Fibonacci ratio sequence, about which you discovered something important in Section 5.3.

Example 10 $x_1 = 1.5$;

thereafter,

$$x_n = \tfrac{1}{2} \left(x_{n-1} + 2/x_{n-1} \right):$$

$$1.5, 1.41666\ldots, 1.414216\ldots, \ldots$$

See Section 4.4. Example 10 is the translation of a sequence found on an old Babylonian clay tablet.

The examples illustrate that sometimes there is a number that the entries of a sequence eventually get close to (a "limit"), and sometimes there is not. For example, the sequences of Examples 1 and 8 have terms that get larger and larger "without bound" and therefore do not get close to any particular number. On the other hand, the sequences of Examples 2 through 5 all approach limits, namely, 1, $\frac{1}{3}$, π, and π, respectively. (See if you can figure out why in each case.) We observed empirically that the sequence R_n in Example 9 appears to approach $(1 + \sqrt{5})/2$. And we observed in Section 4.2 (and prove in Sect. 11.4) that the sequence x_n in Example 10 approaches $\sqrt{2}$. Much less obvious are the facts that the sequence in Example 6 does not approach a limit, whereas the one in Example 7 does approach a limit. (See the Digression in Sect. 4.2.)

7.2 LIMITS

What does it really mean to say that, "if we go far enough out in this sequence, all the terms get close to the number L, the limit?" Here is the formal definition.

Definition   The sequence S *approaches* a number L (or *has L as a limit,* or *converges to L*) provided the following condition is satisfied: For each positive number ϵ, there exists an integer N such that

$$|S_n - L| < \epsilon \qquad \text{whenever } n \geq N.$$

Notice that the vague phrases "close to" and "far enough out" do not appear in the formal definition. Nevertheless, the condition stated in the definition is an expression of the intuitive idea that the terms of the sequence eventually get arbitrarily close to L: No matter how small the number ϵ is chosen, if we go "far enough out" in the sequence—i.e., terms numbered beyond the corresponding N—all subsequent terms S_n will lie within ϵ of L. (Recall that $|S_n - L|$ is just the distance between S_n and L on the number line.)[2]

The significance of this concept for the use of the computer is this: Many of the things we will want to compute are, in fact, limits of sequences, the terms of which may be computed with ease. If we have determined (perhaps by the use of calculus) that an unknown quantity L is the limit of a known—i.e., computable—sequence S, then we can find L to any desired number of decimal places by approximating it by some S_n for a sufficiently large n. For example, if we want five correct decimal places, then we can use S_N, where N is the integer corresponding to $\epsilon = 0.000005$ in the definition. (Why?) In practice, it is usually not so easy to tell what N goes with a given ϵ, even if you know one has to exist, so other means must be devised for determining when a given term of a sequence is adequate as an approximation to the limit. We devote a considerable amount of attention to devising such means.

One point concerning the definition requires special emphasis. The condition for a given number to be a limit of a given sequence is, in reality, *infinitely many* conditions: For *every* positive number ϵ, one must find an appropriate N, and then show that the right inequality is satisfied for *every* integer $n \geq N$. In general, there is no way to reduce this to a finite number of conditions; therefore, there is no way that a computer can *prove* the existence of a limit for a given sequence because it can check only a finite number of things for us. Our usual procedure is to use calculus to show that a limit exists, and then use the computer to compute the limit. Calculus also comes into play in deciding how far out in the sequence we need to go to achieve desired accuracy in the answer.

Convergence of some sequences can be determined directly from the definition. We illustrate this with some examples that will be useful later.

Example 11   The sequence $S_n = 1/n$ converges to 0. This statement is abbreviated

$$\lim_{n \to \infty} \frac{1}{n} = 0.$$

To see this, we have to show that, for any $\epsilon > 0$.

$$\left| \frac{1}{n} - 0 \right| < \epsilon,$$

---

[2] One of the major contributions of the nineteenth century to the development of calculus (due mainly to the German mathematician Karl Weierstrass, 1815–1897) was the careful formulation of limit concepts in terms of elementary properties of the real numbers, such as difference, absolute value, and inequalities—no "moving" variables, infinitely small distances, or other vague terms. (See Grabiner.)

Figure 7.1

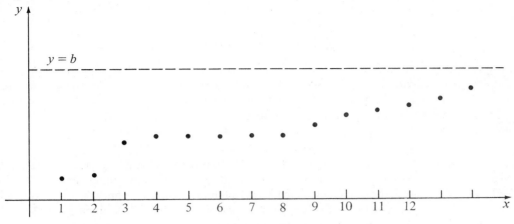

where n is large enough. But $|(1/n) - 0| = |1/n| = 1/n$, and $1/n < \epsilon$ if and only if $n > 1/\epsilon$. (Why?) Hence if N is any integer $> 1/\epsilon$, then $1/n < \epsilon$ for all $n \geq N$.

Example 12 $$\lim_{n \to \infty} \frac{1}{2^n} = 0$$

This should come as no surprise, given the previous example, because $2^n > n$ for positive integers n,[3] so $1/2^n < 1/n$ (Why?), and hence closer to 0. Can you give a formal proof that the limit is 0?

It is often important to be able to show that a sequence converges without having to make a guess at its limit. (If we know the limit already, what is left for the computer to do?) Figure 7.1 illustrates a situation in which we can do just that. The sequence S_n shown there is *increasing* in the sense that $S_1 \leq S_2 \leq S_3 \leq \cdots$. Furthermore, there is a number b with the property that $S_n \leq b$ for every n. Such a sequence is called a *bounded increasing* sequence. Now it would appear that, if new terms of the sequence can never be lower than any earlier terms or higher than $y = b$, then eventually there must be a "leveling off," i.e., convergence to a limit L (which might be b or some smaller number).

Fact A bounded increasing sequence converges.

Although the fact just stated agrees with our intuition (as reinforced by Fig. 7.1), its truth involves a subtlety that is discussed further in Chapter 9.

In Section 7.1, Examples 1, 3, 4, 6, 7, and 8 are increasing, whereas 2, 5, and 9 are not increasing (Verify). Note that some of the increasing sequences are bounded (e.g., Examples 3 and 4), and therefore convergent, and some are not bounded (e.g., Examples 1 and 8). But

---

[3]The fact that $2^n > n$ may be established by using the *binomial theorem*, which tells you how to expand $(x + y)^n$ in powers of x and y. Taking $x = y = 1$, $2^n = (1 + 1)^n = 1 + n +$ other positive terms, which sum is clearly greater than n. The inequality may also be established by induction.

note also that a sequence does not have to be increasing to converge, as Examples 2, 5, and 9 show.

Exercise

1. Formulate definitions for *decreasing* and *bounded decreasing* sequences. Draw a picture like Figure 7.1 to illustrate the fact that a bounded decreasing sequence must converge. Check that Examples 11 and 12 are bounded decreasing sequences.

The next two "lab manual" chapters (8 and 10) ask you to use the computer to examine some additional sequences that are useful later. Chapter 9 need not be taken up at this time unless your calculus course includes a more thorough discussion of sequences at this point, but it explains the reasoning behind the most important properties of sequences and limits. Chapter 11 (also optional) contains proofs of some of the convergence results already noted, or to appear in Chapters 8 and 10.

References

Gaughan; Grabiner (especially Sect. 4); Thomas.

C H A P T E R   8

POWERS AND FACTORIALS [L]

Before continuing the formal study of sequences and their applications, we will observe the convergence of a number of different kinds of sequences. In addition to some rather obvious examples in Section 7.2, you have had occasion to experience "convergence" in at most three situations: (1) approximate computation of the Euler sum

$$\pi^2/6 = 1 + \frac{1}{2^2} + \frac{1}{3^2} + \dots$$

by computing only *one* term of the sequence of partial sums, that for $n = 1000$ (Sects. 4.2 and 5.2); (2) convergence of ratios of Fibonacci numbers to the "golden ratio" (Sects. 4.3 and 5.3), illustrated by computing a fixed number of terms of the sequence; and (3) the

Babylonian computation of square roots (Sects. 4.4 and 5.4), in which there was a programmed decision about when to stop computing terms of the sequence.

In this laboratory session you are asked to do some further experimentation with certain sequences that make some of the later chapters more meaningful. In each part of the session, you have to prepare a short program consisting of a single loop in which a term of a sequence is computed and printed. The loop is preceded by some initialization or input of the appropriate variables. No decisions about interrupting loops are required.

8.1 SEQUENCES OF POWERS

A sequence of numbers in which the ratio of each term to the previous one is a constant r is called a *geometric progression.* Since the first term may be any number b, such a sequence must have the form

$$b, br, br^2, br^3, \ldots .$$

As we see in the next chapter, the constant factor b is irrelevant for questions of convergence, so let's examine sequences of the form

$$T_n = r^n$$

for various values of r. A moment's reflection shows that different limits are possible: if $r = 1$, T is the constant sequence $1, 1, 1, \ldots$; if $r = 100$, its powers grow without bound; if $r = 0.01$, the powers get smaller and smaller. Are there other possibilities?

Write a program with R as an input variable that prints successive powers of R up to the fifteenth power. Note that $R^n = R \cdot R^{n-1}$. Hence, if $T = 1$ before entering the loop in the program, the body of the loop may consist of an assignment that replaces T by R * T and an output statement to print T.

Execute your program with the following values of R:

$$(1)\ 0.1 \qquad (2)\ -0.1 \qquad (3)\ 4 \qquad (4)\ -4$$

So far, it appears that small values of R result in $\lim_{n \to \infty} R^n = 0$, and even moderately large values fail to lead to convergence. Continue execution with the following values of R.

$$(5)\ 0.25 \qquad (6)\ 0.9 \qquad (7)\ 1.1$$

Are you prepared to believe the following assertion? (This is a theorem, the proof of which may be found in Sect. 11.1.)

Fact   If $|r| < 1$, then $\lim_{n \to \infty} r^n = 0$. If $|r| > 1$, then $\lim_{n \to \infty} r^n$ does not exist.

8.2 CAN YOU CROSS THE ROOM?

Consider the following (philosophical) problem: If you want to walk across the room, you first have to walk halfway across the room. You then have to walk half the remaining distance ($1/4$ of the width of the room), then half the remaining distance ($1/8$ of the width), and so on. This means you must traverse an *infinite* number of (progressively smaller) distances.

But you cannot carry out an infinite number of tasks if each requires a finite amount of time. Does that mean you *cannot* cross the room? (If you are having doubts, stand up and try.)[1]

If we take the width of the room as the unit of distance, what the preceding paragraph really suggests is that

$$1 = \frac{1}{2} + \frac{1}{4} + \frac{1}{8} + \frac{1}{16} + \cdots,$$

i.e., it is possible to *sum* the terms of a geometric progression (with $b = \frac{1}{2}$ and $r = \frac{1}{2}$) and get a finite answer (1 in this case).

Write a program to evaluate the first 20 partial sums (terms of the sequence):

$$S_n = B + BR + \cdots + BR^n.$$

Take B and R as input variables, and set S and T equal to B. In the loop, compute T as in Section 8.1 and $S = S + T$; then print S. Execute the program with the following values for B and R, respectively:

$$(1)\ 0.5, 0.5 \qquad (2)\ 0.3, 0.1 \qquad (3)\ \frac{1}{3}, \frac{1}{2} \qquad (4)\ 7, \frac{2}{3}$$

Exercise (2) may suggest that certain sums of geometric progressions are really quite familiar.

WARNING: A fraction such as $\frac{1}{3}$ will not be a valid input to the computer. Use an appropriate decimal approximation instead.

It is not really necessary to use a computer to evaluate sums of geometric progressions because there is an explicit formula for the sum. (The proof is not difficult. You will find it in Sect. 11.2.)

Fact
$$b + br + br^2 + \cdots + br^n = \frac{b(1 - r^{n+1})}{1 - r}$$

In particular, if $|r| < 1$, then

$$\lim_{n \to \infty} (b + br + br^2 + \cdots + br^n) = \frac{b}{1 - r}.$$

Note that the second statement follows from the first by using the fact stated in Section 8.1: The term r^{n+1} has the limit 0 as n becomes large. Check your answers for Exercises (1) through (4) against the formula just stated.

8.3 POWERS VERSUS FACTORIALS

If n is a positive integer, the product of all the positive integers $\leq n$,

$$1 \cdot 2 \cdot 3 \cdots (n - 1)n,$$

---

[1]This illustration is a variant of the Achilles-and-the-tortoise paradox, one of four such paradoxes formulated by Zeno of Elea (ca. 450 B.C.) to show that "motion" is an illusion. See Boyer (1968) for a discussion of all four paradoxes, and the other references for related material.

is called *n factorial,* and is abbreviated *n*!. (The exclamation point is part of the notation, not punctuation for the sentence.) Note that 1! has only one factor: 1! = 1. For reasons of convenience (which will emerge later), we also define 0! = 1. Such products grow rather rapidly as *n* increases. What happens when they are compared with other rapidly growing products, such as powers r^n, with $r > 1$? (See Sect. 8.1.) Let's examine the sequence

$$S_N = \frac{R^N}{N!},$$

where R is taken as an input variable. When $R = 2$, say, and $N \geq 3$, we have

$$S_N = \frac{2 \cdot 2 \cdot 2 \cdots 2}{1 \cdot 2 \cdot 3 \cdots N},$$

and evidently $S_N \leq 2$ after the first two terms, so the sequence certainly does not have arbitrarily large terms. In fact, the factorial in the denominator would appear to grow more rapidly than the power in the numerator (as N grows), but how much more rapidly?

To simplify the computation, let's observe (as we did in Sect. 8.1) that

$$S_N = \frac{R^N}{N!} = \frac{R^{N-1} \cdot R}{(N-1)! \, N} = \frac{S_{N-1} R}{N}.$$

Thus, if we set $S = 1$ to begin with, the body of the loop indexed by N may compute $S = S * R / N$. Write such a program, with $N = 1, 2, \ldots, 30$, and execute it with: (1) $R = 2$, and (2) $R = 10$.

You may try other cases if you wish. Would you like to make a guess about

$$\lim_{n \to \infty} \frac{r^n}{n!}$$

in general? (The correct guess is stated and proved in Chap. 37, where it is first needed.)

References

Boyer (1968); Gardner (1964, 1971).

C H A P T E R 9

USEFUL FACTS ABOUT SEQUENCES [0]

We come now to the hard work that must be done in order to understand computations with sequences. Each of the "properties" that follow expresses an important, but intuitively obvious, *theorem* about limits. We provide *reasons* for each to aid your intuition. For actual proofs, see Thomas or the appropriate sections of your calculus book. We emphasize, however, that it is not necessary for you to *master* the details of proofs about limits of sequences at this time.

Property 1   Only the "tail" of a sequence matters—i.e., any finite number of terms may be discarded, and what is left will converge to L if and only if the original sequence converged to L.

Reason   It makes no difference how large the "N" in the definition (Sect. 7.2) is. If one exists (for a given ϵ), any larger integer will do just as well, and it can be chosen larger than the finite number of terms discarded. If no such N exists (for a given ϵ), then discarding terms will not change anything because, in particular, no very large such N exists.

Remark   Property 1 may be turned around to say that any finite number of terms may be added at the start of a sequence (or inserted at arbitrary places in the sequence) without affecting convergence.

Definition   A set X of numbers is *bounded above* if there is some number b such that $b \geq x$ for every $x \in X$. Similarly, one may define *bounded below*. (Do it!) A set that is bounded above and below is called *bounded*. A sequence S is *bounded* (respectively, *bounded above, bounded below*) if the set $\{S_n\}$ of all its values is bounded (respectively, bounded above, bounded below.)

Property 2   A convergent sequence is bounded.

Reason   The intuitive content of this statement is that if S_n is busy converging to the (fixed) number L, it cannot also be marching off to infinity ($+$ or $-$). To be a little more precise, we may apply the definition of $\lim_{n \to \infty} S_n = L$, taking $\epsilon = 1$, say. A corresponding N must exist so that $|S_n - L| < 1$ for all $n \geq N$. In other words, the tail of the sequence from N on is bounded above by $L + 1$ and below by $L - 1$. This accounts for all but the first $N - 1$ values of S, and clearly any finite set is bounded. Finally, it is easy to see that the union of two bounded sets (in this case $\{S_1, \ldots, S_{N-1}\}$ and the tail) is bounded.

Remark Property 2 is used usually in the alternative form: An unbounded sequence does not converge (see Examples 7.1 and 7.8). Note that a bounded sequence need not converge—e.g., the sequence 1, 0, 1, 0, 1, 0, ... is bounded but has no limit. (Why?)

Property 3 $\lim_{n \to \infty} S_n = L$ if and only if $\lim_{n \to \infty} |S_n - L| = 0$.

Reason For clarity, let's set $T_n = |S_n - L|$ for each n. The second limit statement has to do with the sequence T. The crucial quantity to be examined in deciding whether or not $\lim_{n \to \infty} T_n = 0$ is

$$
\begin{aligned}
|T_n - 0| &= ||S_n - L| - 0| \\
&= ||S_n - L|| \\
&= |S_n - L|, \qquad \text{(Why?)}
\end{aligned}
$$

i.e., exactly the same quantity involved in the definition of $\lim_{n \to \infty} S_n = L$. (Can you construct a formal proof?)

Remark The significance of Property 3 is that all convergence questions can be turned into questions about convergence of (nonnegative) sequences *to zero*.

Example 7.2 *(continued)* We asserted earlier that

$$
\lim_{n \to \infty} \left[1 + \frac{(-1)^n}{2^n} \right] = 1.
$$

By Property 3, in order to *prove* this, we could show instead that

$$
\left| \left[1 + \frac{(-1)^n}{2^n} \right] - 1 \right|
$$

converges to zero. Simplification of this expression leads to $1/2^n$ (check it!), and we saw that

$$
\lim_{n \to \infty} \frac{1}{2^n} = 0
$$

in Example 7.12.

Property 4 If $S_n = R_n T_n$ for each n, where R is a bounded sequence (not necessarily convergent!) and T converges to zero, then S also converges to zero.

Reason If R is a bounded sequence, so is $|R|$; i.e., $|R_n| \leq b$ for some number b. Then $|S_n| = |R_n||T_n| \leq b|T_n|$. By Property 3 (with $L = 0$), $\lim_{n \to \infty} T_n = 0$ if and only if $\lim_{n \to \infty} |T_n| = 0$. Because the numbers $|T_n|$ eventually get close to zero, so do the numbers $|S_n|$.

Remark This property is a tool for simplifying complicated problems. For example,

$$\lim_{n \to \infty} \frac{[1 + (-1)^n]^{1/n}}{n} = 0,$$

because $R_n = [1 + (-1)^n]^{1/n}$ is bounded (by what?) and $T_n = 1/n$ converges to zero.

Properties 5 If $\lim_{n \to \infty} S_n = L$ and $\lim_{n \to \infty} T_n = M$, then

a. $\lim_{n \to \infty} (S_n + T_n) = L + M$.

b. $\lim_{n \to \infty} (S_n - T_n) = L - M$.

c. $\lim_{n \to \infty} (S_n T_n) = LM$.

Furthermore, if $T_n \neq 0$ for every n, and $M \neq 0$, then

d. $\lim_{n \to \infty} (S_n / T_n) = L/M$.

Reason It should come as no surprise that, if S_n is close to L and T_n is close to M, $S_n + T_n$ should be close to $L + M$, and similarly for the other properties. However, formal proofs of these statements involve some computations that are a little tricky. If you have studied the proofs for the corresponding properties for limits of *functions,* exactly the same tricks work.

Example 7.9 (continued) Recall from the discussion in Section 5.3 that the Fibonacci ratio sequence R_n satisfies

$$R_{n+1} = 1 + \frac{1}{R_n}.$$

Now if $S_n = R_{n+1}$, then the sequence S is obtained from R by discarding the first term. Hence, by Property 1,

$$\lim_{n \to \infty} R_{n+1} = \lim_{n \to \infty} R_n.$$

If there is such a limit, say, r, then

$$
\begin{aligned}
r = \lim_{n \to \infty} R_{n+1} &= \lim_{n \to \infty} \left(1 + \frac{1}{R_n} \right) \\
&= 1 + \frac{1}{\lim\limits_{n \to \infty} R_n} \qquad \text{(Why?)} \\
&= 1 + \frac{1}{r}.
\end{aligned}
$$

This was the equation that led to the conclusion that $r = (1 + \sqrt{5})/2$.[1]

---

[1] An elementary proof that this "golden ratio" really is the limit may be found in "Beginner's Corner," *Fibonacci Quarterly* **1**, No. 3 (1963), 53–55.

A word of caution is in order concerning the use of Properties 5 (a) through (d). Suppose we consider the Fibonacci sequence F_n (1, 1, 2, 3, 5, . . .), and write

$$\lim_{n \to \infty} F_n = f.$$

The same reasoning that we just used for the ratio sequence, applied to the identity

$$F_n = F_{n-1} + F_{n-2},$$

leads to the conclusion

$$f = f + f = 2f,$$

or, apparently, $f = 0$, which is absurd. What went wrong?

Definition A sequence S is *increasing* if

$$S_1 \le S_2 \le S_3 \le \cdots$$

(i.e., $S_n \le S_{n+1}$ for every n). Similarly, one defines *decreasing*. (Do it!) A sequence that is either increasing or decreasing is called *monotone*.

Property 6 A bounded monotone sequence is convergent.

The reasoning behind Property 6 was discussed in Section 7.2, at least for the increasing case. However, it is not really obvious that the "leveling off" that necessarily occurs actually leads to convergence to a real number, until one takes into consideration the most subtle property of real numbers: the *completeness axiom*. This axiom has a number of different formulations, one of which is precisely Property 6. Thus this property is different from the others we are listing, in that it is not really a theorem to be proved, but rather, a fundamental *assumption* about real numbers. It is basically this property that separates real numbers from rational numbers (quotients of integers) and makes the former a more useful number system with which to work. Note that the sequence of Example 7.4 is bounded above (by 4, say) and increasing. Each of its terms is a rational number. (Why?) But it does not converge to a rational number, since π is irrational (a nontrivial fact). Property 6 asserts that this sequence *does* converge to some real number.

Property 7 If $R_n \le S_n \le T_n$ for every n, and

$$\lim_{n \to \infty} R_n = \lim_{n \to \infty} T_n = L,$$

then the middle sequence S also converges, and to the same limit:

$$\lim_{n \to \infty} S_n = L.$$

Reason Intuitively, as R_n and T_n both get close to L, they also get close to each other. S_n is caught in between, so it gets close to both R_n and T_n and joins them in converging to L. More formally, given $\epsilon > 0$, we can go far enough out in both the R and T sequences so that R_n and T_n differ from L by less than ϵ (for all subsequent values of n). That is,

$$L - \epsilon < R_n < L + \epsilon$$

and

$$L - \epsilon < T_n < L + \epsilon.$$

Combining these inequalities with the one stated in Property 7, we have

$$L - \epsilon < R_n \leq S_n \leq T_n < L + \epsilon.$$

Hence S_n also lies within ϵ of L for all subsequent values of n.

Remark   Property 7 is variously called the "Pinching Theorem," "Sandwich Theorem," or "Squeeze Theorem."

Example 7.12 *(continued)*   Given the fact that $\lim_{n \to \infty} 1/n = 0$ (Example 7.11), the justification for stating that $\lim_{n \to \infty} 1/2^n = 0$ can be based on the Pinching Theorem. Because $0 < 1/2^n < 1/n$ for each positive integer n, we may set $R_n = 0$, $T_n = 1/n$, and $S_n = 1/2^n$. The desired conclusion follows immediately from Property 7.

   More sophisticated examples of the use of the limit properties are presented in Chapter 11, after you have had an opportunity to experiment with some more sequences in your next laboratory session. In particular, Chapter 11 provides proofs of some of the claims we have made thus far but have not substantiated.

Remark   Both Properties 6 and 7 have hypotheses involving inequalities that must hold *for all n*. It will often be convenient to combine these statements with Property 1 to conclude that it is sufficient to have the necessary inequalities satisfied "in the tail"—i.e., after discarding a finite number of terms. Thus a bounded sequence that is *eventually* monotone converges, and a sequence eventually sandwiched between convergent sequences with a common limit also converges.

C H A P T E R   1 0
SEQUENCES AND DERIVATIVES [L]

This laboratory session continues the program of Chapter 8 for experimenting with convergence while studying the mathematical properties of sequences. Two quite different types of exercises are presented here, but in each case the programming required is just like that in Chapter 8: a single fixed-length loop to compute a certain number of terms of a particular sequence.

10.1 HOW (NOT) TO COMPUTE DERIVATIVES

By now you have seen in your calculus course some expressions such as

$$\lim_{x \to a} F(x)$$

or

$$\lim_{\Delta x \to 0} \frac{f(x_1 + \Delta x) - f(x_1)}{\Delta x}.$$

At this stage of the game, such expressions are more easily evaluated with pencil and paper than by computer. As a general rule, you are not encouraged to use the computer for problems that are more appropriately done some other way, but we bend that rule a little for this exercise.

The notation

$$\lim_{x \to a} F(x)$$

expresses something about the values of a function F for *all* numbers x, except a, near a given number a, whether or not F has a value at a. In particular, if F is a difference quotient (of another function f), and we replace x by Δx and a by 0, so that

$$\lim_{\Delta x \to 0} F(\Delta x) = \lim_{\Delta x \to 0} \frac{f(x_1 + \Delta x) - f(x_1)}{\Delta x} = f'(x_1),$$

then the limit exists whenever f is differentiable at x_1, even though the limit is *not* a value of F at 0. The computer cannot help us examine $F(\Delta x)$ for *every* Δx near 0, but we can consider a sequence of numbers converging to 0, such as

$$1, \frac{1}{2}, \frac{1}{3}, \ldots$$

and the corresponding sequence

$$F(1), F\left(\frac{1}{2}\right), F\left(\frac{1}{3}\right), \ldots.$$

It is a fact that, if $\lim_{\Delta x \to 0} F(\Delta x)$ exists, then for every sequence $S_1, S_2, \ldots$ such that $\lim_{n \to \infty} S_n = 0$,

$$\lim_{n \to \infty} F(S_n) = \lim_{\Delta x \to 0} F(\Delta x).$$

Thus, *if* existence of the limit is known, it can be computed as the limit of a sequence.

Before plunging ahead with a computation, we should take advantage of the fact that any sequence S_n with limit 0 will work. $S_n = 1/n$ is one possibility, but it happens not to be a very good choice because we would like to get as close to the limit as we can, as quickly as possible. For this purpose, a much faster converging sequence such as $S_n = 1/2^n$ is preferable.

Thus, given a function f and a fixed number x at which it is differentiable, we have

$$f'(x) = \lim_{n \to 0} \frac{f(x + 1/2^n) - f(x)}{1/2^n}$$
$$= \lim_{n \to \infty} 2^n [f(x + 1/2^n) - f(x)]$$
$$= \lim_{n \to \infty} Q_n,$$

where Q_n is defined by the preceding line.

Exercises

1. Using a function definition statement for f and an input statement for x, write a program to compute and print 15 terms of the sequence Q_n. Rather than compute 2^n by exponentiation, you might set $K = 1$ before entering the loop, and then replace K by $2 \cdot K$ each time through. The nth time through, K will have the value 2^n. (Why?)

2. Now let's try out the program with a function for which you already know the answers:

$$f(x) = x^3 - 3x + 4.$$

Of course, $f'(x) = 3x^2 - 3$. Run your program with $x = 1$ and then with $x = 4$. Do you get apparent convergence to the right numbers, at least approximately?

3. Next, let's try the program with a function for which you may not yet know the derivative:

$$f(x) = \sin x.$$

Use $x = 0.7$ and then $x = -1.3$. (The derivative of the sine function happens to be the cosine function. Check your answers against $COS(0.7)$ and $COS(-1.3)$, obtained by direct inquiry, a one-line program, or your calculator.)

10.2 SOME MILDLY SURPRISING LIMITS

For no particularly obvious reason, let's consider the sequence S whose nth term is given by

$$S_n = \left(1 + \frac{1}{2^{2n}}\right)^{2^n}.$$

This is an example of a situation where there is a simple formula for the typical term, but it may be difficult to guess what the limit (if any) should be (unlike Example 7.2, for instance). The expression $1 + 2^{-2n}$ converges to 1, of course, but it is being raised to larger and larger powers, and it is always greater than 1. Does S_n converge to 1? Are there arbitrarily large terms of the sequence? Or is there some number > 1 that is the limit?

Exercises

1. Write a single-loop program to print the values of S_n, where $n = 1, 2, \ldots, 15$. Here is a hint to simplify the programming: As in Section 10.1, set $K = 1$ before entering the loop and let $2K$ replace K each time through. If N is the loop index, $K = 2^N$, so

$$S_N = \left[1 + \frac{1}{(K \cdot K)} \right]^K.$$

The right-hand member of that equation can be used in an assignment statement to define S with a minimum of exponentiation.

Now run the program. Do the results suggest that $\lim_{n \to \infty} S_n = 1$?

2. Repeat the previous exercise with a slightly different sequence:

$$S_n = \left(1 + \frac{1}{2^n} \right)^{2^n}.$$

Note that the only difference is in the exponent in the denominator of the second term. Thus the only change required in your previous program is to make the assignment statement read:

$$\text{LET} \quad S = (1 + 1/K)\uparrow K$$

Now run the program again. What number appears to be the limit? Is the answer a little surprising?

3. This time have the program print 15 values of the sequence

$$S_n = \left(1 - \frac{1}{2^{2n}} \right)^{2^n}.$$

Note that this is just like exercise (1) except that the plus sign is replaced by a minus sign. Hence each quantity raised to a power is slightly *less* than 1. Again, you are raising numbers very close to 1 to very large powers. What do you think the limit (if any) will be? Run the program. What does the limit appear to be?

4. Finally, do the same with the sequence

$$S_n = \left(1 - \frac{1}{2^n} \right)^{2^n}.$$

which is just like part (2) except for the sign change. Do you expect an answer like that for (2) or like that for (3)? Run the program. Are you a little surprised again, or is it just what you expected?

NOTE: If you have had occasion to study natural logarithms and the exponential function, perhaps you were not surprised at all, and indeed, already know why the sequences in (2) and (4) behave differently from those in (1) and (3). If not, the mystery will be dispelled later in your calculus course and in Chapter 45 of this book. The number that is the limit in part (2) is universally known as e, and the limit in part (4) is its reciprocal.

C H A P T E R   1 1

APPLICATIONS OF
LIMIT THEOREMS [0]

We looked at some basic ideas about sequences in the abstract in Chapters 7 and 9 and examined some specific sequences in Chapters 4, 5, 8, and 10. The time has come to apply the abstract ideas to some of the specific cases and thereby establish some important facts. Both the facts and the techniques used for establishing them are used in later, optional chapters. This is especially true of the first three sections in which we consider geometric progressions and their sums and the relationships between limits of sequences and limits of functions. Less important for immediate use, but included to satisfy your burning curiosity, is the final section in which we verify the Babylonian square root method.

11.1 CASE STUDY: SEQUENCES OF POWERS (GEOMETRIC PROGRESSIONS)

We have already noted in Section 8.1 that the convergence of a power sequence

$$T_n = r^n$$

depends primarily on the size of r, specifically, on whether $|r| < 1$, $|r| = 1$, or $|r| > 1$. The most important property of power sequences is expressed as follows:

Theorem 1   If r is a real number, and $|r| < 1$, then

$$\lim_{n \to \infty} r^n = 0.$$

Proof   First note that in order to show the limit is 0, it is sufficient to show that

$$\lim_{n \to \infty} |r^n| = \lim_{n \to \infty} |r|^n = 0. \qquad \text{(Property 9.3 with } L = 0\text{)}$$

Thus we may suppose r is nonnegative to begin with. Of course, if $r = 0$, the result is obvious. (Why?) Hence we may assume

$$0 < r < 1.$$

Now $r^n > 0$, so if we multiply both sides of $r < 1$ by r^n, we get

$$r^{n+1} < r^n.$$

This shows that the sequence of powers is decreasing. It is also bounded below, by 0, and hence converges (Property 9.6) to some unknown limit; i.e.

$$L = \lim_{n \to \infty} r^n.$$

We also have

$$L = \lim_{n \to \infty} r^{n+1}. \qquad \text{(Property 9.1)}$$

We may take limits on both sides of the equality

$$r^{n+1} = r \cdot r^n$$

and conclude that

$$L = rL. \qquad \text{(Property 9.5 [c])}$$

Hence $(1 - r)L = 0$, so either $1 - r = 0$ or $L = 0$. (Why?) Because $r \neq 1$, we must conclude that $L = 0$. Q.E.D.

Exercises

1. What can you say about $\lim_{n \to \infty} T_n$ in the cases $r = 1$ and $r = -1$?

2. Show that the sequence T is unbounded (and therefore not convergent) when $|r| > 1$.

    HINT: When $|r| > 1$, $|1/r| < 1$, and $(1/r)^n = 1/r^n$.

11.2 CASE STUDY: SUMS OF GEOMETRIC PROGRESSIONS

As we observed in Section 8.2, one of the ancient paradoxes of Zeno can be resolved by showing that

$$\frac{1}{2} + \frac{1}{4} + \frac{1}{8} + \frac{1}{16} + \cdots$$

adds up to 1, i.e., the limit of the sequence of partial sums is 1. The earliest known computation of the sum of a geometric progression was carried out some two centuries after Zeno by the greatest mathematician of antiquity, Archimedes of Syracuse (287?–212 B.C.).[1] Archimedes showed by indirect arguments that

$$1 + \frac{1}{4} + \frac{1}{4^2} + \cdots + \frac{1}{4^n} + \cdots$$

must be neither more nor less than ⁴⁄₃. He couldn't actually say the *sum is* ⁴⁄₃ because he and his contemporaries didn't believe in infinite processes. Incidentally, Archimedes was not concerned with settling Zeno's paradoxes, but rather, was computing the area of a parabolic segment, a problem you will soon learn to associate with integral calculus.

Both of the sums mentioned previously, and all the ones you computed in Section 8.2, are special cases of the following:

---

[1] See Boyer (1968), Chapter 8.

Theorem 2 If b and r are real numbers, with $|r| < 1$, and

$$S_n = b + br + \cdots + br^n, \quad \text{for } n = 1, 2, \ldots,$$

then

$$\lim_{n \to \infty} S_n = \frac{b}{1 - r}.$$

Proof We have

$$S_n = b + br + br^2 + \cdots + br^{n-1} + br^n$$

given, and

$$rS_n = br + br^2 + \cdots + br^{n-1} + br^n + br^{n+1},$$

upon multiplying both sides by r. Subtracting the second equation from the first, we get

$$(1 - r)S_n = b - br^{n+1}.$$

Hence

$$S_n = \frac{b(1 - r^{n+1})}{1 - r}, \quad \text{since } r \neq 1.$$

Then

$$\lim_{n \to \infty} S_n = \frac{b}{1 - r} \lim_{n \to \infty} (1 - r^{n+1}) \quad \text{(Why?)}$$

$$= \frac{b}{1 - r} (1 - \lim_{n \to \infty} r^{n+1}) \quad \text{(Why?)}$$

$$= \frac{b}{1 - r}, \quad \text{(Why?)}$$

Q.E.D.

Exercises

3. If S_n is defined as shown previously, what can you say about $\lim_{n \to \infty} S_n$ in the cases $r = 1$ and $r = -1$?

4. If $r > 1$, show that the sequence S is unbounded, and hence does not converge. Can you modify the argument to show that S is unbounded in the case $r < -1$?

5. Use the Theorem to compute each of the sums from Section 8.2. Do your results agree with the computer results?

6. Write the repeating decimal

$$0.197219721972\ldots$$

as a fraction (quotient of integers).

HINT: Consider a geometric progression with $r = 10^{-4}$.

11.3 CASE STUDY: LIMITS OF FUNCTIONS

In Section 10.1 we mentioned the connection between limits of sequences and limits of functions (including derivatives as limits of difference quotients). Intuitively,

$$\lim_{x \to a} f(x) = L$$

means that all the values of f are close to L at numbers x in the domain that are sufficiently close to a (except possibly for a itself). In particular, if $a_1, a_2, \ldots$ is a sequence converging to a, then eventually—i.e., for large enough n—all the numbers a_n are close to a, so the numbers $f(a_n)$ must be close to L, i.e., $\lim_{n \to \infty} f(a_n)$ should also be L. Here is a more formal treatment:

Theorem 3   If

$$\lim_{x \to a} f(x) = L,$$

then, for every sequence $\{a_n\}$ in the domain of f such that $\lim_{n \to \infty} a_n = a$ (but with no $a_n = a$), we have

$$\lim_{n \to \infty} f(a_n) = L.$$

NOTE: The converse of this statement is also true but is of no use to us in this course.

Proof   The hypothesis asserting the existence of the limit L may be expanded to read: For every $\epsilon > 0$, there exists a $\delta > 0$ such that

$$|f(x) - L| < \epsilon \qquad \text{whenever} \qquad 0 < |x - a| < \delta \qquad (11.1)$$

(and x is in the domain of f). To establish the conclusion, we have to show that, for every $\epsilon > 0$, there exists an integer N such that

$$|f(a_n) - L| < \epsilon \qquad \text{whenever} \qquad n \geq N. \qquad (11.2)$$

Given an $\epsilon > 0$, let $\delta > 0$ be selected satisfying condition (11.1). Because $\lim_{n \to \infty} a_n = a$, we know that there exists an integer N such that

$$|a_n - a| < \delta \qquad \text{whenever} \qquad n \geq N. \qquad (11.3)$$

(δ is, after all, one of the positive numbers to which we may apply the definition of Sect. 7.2.) Then, for $n \geq N$, the numbers $x = a_n$ satisfy the "whenever" of condition (11.1); hence $|f(a_n) - L| < \epsilon$ as required for establishing the truth of inequality (11.2). Q.E.D.

11.4 CASE STUDY: BABYLONIAN SQUARE ROOTS

In Section 4.4 we explained how the ancient Babylonians computed $\sqrt{a}$ by successive averaging of x and a/x. Your experience with Section 5.4 may have convinced you that the method actually works, and quite rapidly. But *why* does it work? Close examination of the discussion in Chapter 4 reveals that we fell far short of a *proof* that $\lim_{n \to \infty} x_n = \sqrt{a}$. That deficiency may now be remedied by completing the series of exercises that follows.

For convenience, the Babylonian method is described in terms of *two* sequences of numbers: Let a and x_1 be arbitrary positive numbers, and $y_1 = a/x_1$. Thereafter, the sequences $\{x_n\}$ and $\{y_n\}$ are defined recursively by

$$x_{n+1} = \frac{x_n + y_n}{2}$$

and

$$y_{n+1} = \frac{a}{x_{n+1}}.$$

Exercises

7. Show that all the ordered pairs (x_n, y_n) and (y_n, x_n) are points on the graph of $y = a/x$, as is the pair $(\sqrt{a}, \sqrt{a})$ (see Fig. 11.1).

Figure 11.1

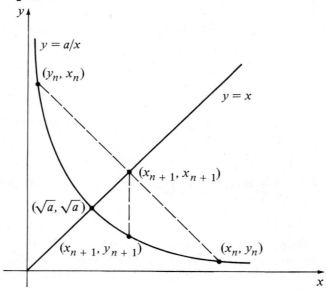

8. Show that the midpoint of the line segment from (x_n, y_n) to (y_n, x_n) is (x_{n+1}, x_{n+1}).

9. Figure 11.1 suggests that $x_{n+1} \geq \sqrt{a}$. Show that this is true. Conclude that if x_1 is to the right or to the left of $\sqrt{a}$, then x_2 and all subsequent x_n's must be to the right.

HINT: Let $g(x) = \frac{1}{2}[x + (a/x)]$. Another way to state the problem is: If $0 < x < \sqrt{a}$, then $g(x) > \sqrt{a}$. Show that $1 - (x/\sqrt{a})$ and $(a/x) - \sqrt{a}$ are both positive and their product is $2[g(x) - \sqrt{a}]$.

10. Show that, for $n \geq 2$, $y_n \leq \sqrt{a}$, $x_{n+1} \leq x_n$, and $y_{n+1} \geq y_n$. Conclude that both sequences are bounded and monotone (after $n = 1$) and hence convergent.

11. Show that, for each n,

$$| x_{n+1} - y_{n+1} | \leq \frac{1}{2^n} | x_1 - y_1 |,$$

and hence that $\lim_{n \to \infty} | x_{n+1} - y_{n+1} | = 0$.

12. Conclude that both sequences converge to $\sqrt{a}$.

HINT: Apply the Pinching Theorem (Property 9.7) with the middle sequence *constant* at $\sqrt{a}$.

C H A P T E R 1 2

AN HISTORICAL PIECE OF PI [R]

As a result of your previous training in mathematics, you are probably aware that the ratio of the circumference of any circle to its diameter is a certain constant, universally known as π. You may also have done some computations by using an approximate value for π, such as 3.14 or $^{22}/_7$. In Section 6.4 we suggested the use of a better approximation, 3.14159. A still better approximation is

$$3.14159\ 26535\ 89793\ 23846\ 26433\ 83279\ 50288\ 41972.$$

Nevertheless, this is *only* an approximation because π is an *irrational* number and therefore has an infinite, nonrepeating decimal expansion.[1]

It has been known since antiquity that the circumference-to-diameter ratio is the same for all circles,[2] but it hasn't always been "known" (in the sense of a suitable approximation) what that ratio is. One of the best ancient approximations, $^{256}/_{81}$ (or about 3.1604), was found in the Rhind Papyrus, which was written in Egypt around 1700 B.C. During the same period the value 3 was in common use in the Mesopotamian (or Babylonian) civilization. The Hebrews apparently adopted this convention from their Semitic ancestors (also Mesopotam-

---

[1]Exercise 6 in Chapter 11 suggests a method for showing that any repeating decimal expansion represents a rational number, i.e., a quotient of integers.

[2]Incidentally, do you know how to prove that?

ian), for we have the following account, dating from the tenth century B.C. of some of the bronze work done by Hiram of Tyre when Solomon was building the temple of Jerusalem:

> Then he made the molten sea; it was round, ten cubits from brim to brim, and five cubits high, and a line of thirty cubits measured its circumference.[3]

On the other hand, a Mesopotamian clay tablet unearthed in 1936 showed that the approximation 25/8 (equal to 3.125) was also known in the Old Babylonian period, which is at least as good as the Egyptian approximation.

Throughout the ages, various approximations to π have been obtained by individuals who thought they were solving the classical problem of "squaring the circle"—i.e., giving a ruler-and-compass construction of the side of a square whose area is equal to the area of a given circle. Such a construction is equivalent to constructing a segment of length π (given a segment of unit length), and this was finally proved impossible in 1882. Because each such (proposed) construction implies a certain value for π, the usual method for proving circle-squarers wrong was to compute more decimal places in the expansion of π and thereby show that the constructed value was not exact.

But crude determinations of π have persisted, and perhaps the most outrageous example in modern times was an attempt to establish π by legislation:

> Be it enacted by the General Assembly of the State of Indiana: It has been found that a circular area is to the square on a line equal to the quadrant of the circumference, as the area of an equilateral rectangle is to the square on one side. . . .

This text is from Section I of Bill No. 246 of the Indiana State Legislature of 1897, and the bill actually passed the House. However, in spite of support for the measure by the Superintendent of Public Instruction, the Senate shelved the bill after it was ridiculed in the newspapers. The story is often retold, erroneously, as a misguided attempt by a kindly school official to make things easier for students by setting the value of π at 3 (the Biblical value), but in fact, if you look closely at the text of the legislation you will see that it implies $\pi = 4$!

Not all ancient approximations to π were as crude as the ones we have mentioned. In fact, the frequently used values of 3.14 and 22/7 date back to Archimedes (ca. 240 B.C.). Accurate computation of π is really no mystery because this number is the measure of the circumference of a circle of diameter 1, and we can approximate that circumference as closely as we please by perimeters of regular polygons, either inscribed or circumscribed. Archimedes did just that: By considering polygons with 96 sides, he showed that π must lie between 223/71 (about 3.1408) and 22/7 (about 3.1429). About four hundred years later, the astronomer Ptolemy of Alexandria computed the perimeter of a polygon with 360 sides to obtain the approximation 377/120 = 3.141666. . . .

Mathematicians of other ancient civilizations also attacked the problem. In the Chinese *Chui-chang suan-shu* (*Nine Chapters on the Mathematical Art*) of about 250 B.C., π is taken to be 3, but in the early part of the Christian era, other values used in China included $\sqrt{10}$ (about 3.162), 99/29 (about 3.172), and 142/45 (3.155 . . .). In the third century A.D., Archimedes'

---

[3]I Kings 7:23 or II Chronicles 4:2. From the Revised Standard Version of the Bible, copyrighted 1946, 1952 © 1971, 1973.

Table 12.1 Computations of π by the Method of Perimeters

Date	Name	No. of computed decimal places	No. of sides of polygon
Ca. 240 B.C.	Archimedes of Syracuse (Greek)	2	$6 \cdot 2^4 = 96$
Ca. 150 A.D.	Ptolemy of Alexandria (Greek)	4	360
Ca. 270	Liu Hui (Chinese)	5	$6 \cdot 2^9 = 3072$
Ca. 480	Tsu Ch'ung-chih (Chinese)	6	?
1430	al-Kashi of Samarkand (Arab)	16	?
1579	François Viète (French)	9	$6 \cdot 2^{16} = 393{,}216$
1593	Adriaen van Roomen (Belgian)	15	$2^{30} \ (>10^9)$
1596	Ludolph van Ceulen (German)	20	$60 \cdot 2^{33} \ (>5 \cdot 10^{11})$
1610	Ludolph van Ceulen (German)	35	$2^{62} \ (>4 \cdot 10^{18})$
1621	Willebrord Snell (Dutch)	35	$2^{30} \ (>10^9)$

method was rediscovered by Liu Hui, who extended the computation to a 3072-sided polygon to obtain $\pi = 3.14159$. Two hundred years later, Tsu Ch'ung-chih obtained the approximation $^{335}/_{113}$ (about 3.1415929) by an unknown method, but he also wrote that π lay between 3.1415926 and 3.1415927, suggesting an extensive computation via perimeters. This degree of accuracy was not equaled in the Western world for another thousand years.

In India, the value $^{3927}/_{1250} = 3.1416$ appeared in the *Paulisha Siddhānta* of about 380 A.D., but the work may have been based on the writings of the astrologer Paul of Alexandria. Similar approximations appear in the work of Āryabhata (ca. 530 A.D.) and Bhāskara (ca. 1150 A.D.), but it is not known whether their results were copied from earlier sources or computed afresh, perhaps from a polygon of 384 sides. Similarly, al-Khowarizmi at Baghdad[4] used a value of 3.1416 (ca. 825 A.D.), but this may have been copied from some Indian source.

Table 12.1 lists some (not all) of the major advances in the computation of π by Archimedes' method, which will be the subject of Chapter 13 because it is another good example of the use of sequences. Such computations reached their peak in 1610, when Ludolph van Ceulen (Germany) achieved the correct value of π to 35 decimal places. This computation took many years and required the determination of the perimeter of a polygon with more than four quintillion sides. The result was so remarkable that it was engraved on his tombstone when he died (the same year) at the age of 70.

In 1621 the Dutch physicist Willebrord Snell worked out a refinement of the perimeter method that allowed determination of a closer pair of bounds for π from any given pair (such as a pair of inscribed and circumscribed perimeters). With this refinement he was able to compute van Ceulen's 35-place result, using polygons with only 2^{30} (still more than a billion) sides. In 1654 another Dutch physicist, Christiaan Huygens, gave the first correct proof of Snell's method, but Huygens' approach to the problem was about to become obsolete with the discovery of calculus methods (in particular, sums of infinite series) on the horizon.

A new era in the computation of π was opened in 1671 by the Scottish mathematician James Gregory, who showed that the radian measure θ of an angle (between $-90°$ and $90°$)

---

[4]See Chapter 3.

is given by

$$\theta = \tan \theta - \frac{\tan^3 \theta}{3} + \frac{\tan^5 \theta}{5} - \frac{\tan^7 \theta}{7} + \cdots. \tag{12.1}$$

The right-hand member of (12.1) is the sum of an infinite series, which, as we have seen, means the limiting value of the sequence of partial sums. One special case of this formula (discovered independently by Leibniz in 1674 and now named after him) is

$$\frac{\pi}{4} = 1 - \frac{1}{3} + \frac{1}{5} - \frac{1}{7} + \cdots. \tag{12.2}$$

This yields a remarkably simple formula for π, but one that is not actually very useful because the convergence is very slow and one must therefore compute the partial sum of a very large number of terms to get even a small number of decimal places of π. The trick in using Gregory's formula effectively is to select angles θ with a known relationship to π such that $\tan \theta$ is as small as possible. Then the increasing powers of $\tan \theta$ will decrease very rapidly, and convergence will also be rapid. (We have much more to say about convergence of infinite series in later chapters.) The first such computation of π was carried out in 1699 by Abraham Sharp in England. Sharp used the angle $\pi/6$, whose tangent is $1/\sqrt{3}$, and correctly computed π to 71 decimal places, more than double the number that had been achieved by van Ceulen.

One complication that arises in using an irrational tangent $(1/\sqrt{3})$ is that its value must be known to a large number of decimal places. In 1706, John Machin (also of England) showed that

$$\frac{\pi}{4} = 4 \arctan \frac{1}{5} - \arctan \frac{1}{239} \tag{12.3}$$

("arctan" means "angle whose tangent is") and used Gregory's series to compute each of the two arctangents, thereby computing π correctly to 100 places. Most extensive computations of π, including those done on computers since World War II, have been done with two-term or three-term arctangent formulas that are similar to Machin's formula. Table 12.2 summarizes the progress in the computation of π, using infinite-series methods, unassisted by computer. In each case I have indicated the number of decimal places subsequently found to be correct (which in many cases was somewhat smaller than the number computed). For example, it was believed that William Shanks had computed 707 correct decimal places in 1873, until D. F. Ferguson showed in 1947 that there were mistakes starting with the 528th place. He also computed the correct value to 710 places. The same year, J. W. Wrench, Jr., published a more extensive computation, but Ferguson found mistakes in it, too. Finally, Ferguson and Wrench jointly published 808 correct decimal places after each had done the computation with a different formula (Wrench used Machin's formula) and each had checked the other's work.

Before we move into the final stage of our saga, the computer era, we should comment briefly on the Japanese entries in Table 12.2. The late seventeenth century saw the emergence of a Japanese version of integral calculus (at about the time of Newton and Leibniz), traditionally attributed (on rather shaky evidence) to Seki Kowa, the "Japanese Newton." The subject was developed by the first and second generation of Seki's students, but actual attributions of specific results are difficult because the subject was treated as a

Table 12.2 Computations of π by Infinite Series Methods

Date	Name	No. of correct decimal places
1699	Abraham Sharp (England)	71
1706	John Machin (England)	100
1719	Thomas Fautet de Lagny (France)	112
1722	Takebe Kenkō (Japan)	41
1730	Matsunaga Ryōhitsu (Japan)	50
1794	Georg von Vega (Austria)	136
1841	William Rutherford (England)	152
1844	Zacharias Dahse (Germany)	200
1847	Thomas Clausen (Germany)	248
1853	William Rutherford (England)	440
1873	William Shanks (England)	527
1947	D. F. Ferguson (England)	710
1947	J. W. Wrench, Jr., and L. B. Smith (U.S.)	722
1948	Ferguson and Wrench	808
1956	Wrench and Smith	1,157

"mystery" and largely kept secret. It is also difficult to tell whether European developments had any effect on the Japanese because intellectual connections with Europe existed, but were few in number. In any case, the Japanese computed areas of circles by using inscribed and circumscribed rectangular polygons (a method that is similar to the one you will see later in your calculus course) and derived various infinite series, including some that led to highly accurate computations of π.

As we have seen, the ultimate achievements in the use of the perimeter method were accomplished by van Ceulen, Snell, and Huygens shortly before the method itself was to become obsolete. The same fate awaited the accomplishments of Ferguson and Wrench, not because of the invention of a better mathematical method, but rather, because of a better computation device. (Note that due to the accelerating pace of new knowledge generally, a "short" time in the seventeenth century might mean a generation, whereas in the twentieth it is more likely to mean a year.) Within a year of the Ferguson–Wrench publication of an 808-place value of π, an already obsolete ENIAC computer at the Army's Ballistics Research Lab (now a museum piece at the Smithsonian) was allowed to run 70 hours to grind out more that 2000 places of π. But that was just a baby step in the computer age, as Table 12.3 shows. The insatiable desire to know the unknown, no longer fueled by the need to disprove attempted circle-squarings, has led to machine computation of a million decimal places for π, and an effort is under way to extend that to 15 million (see Miel). Actually, there are good reasons for wanting this information, as it provides evidence bearing on some interesting statistical questions about the distribution of digits in the expansion of an important nonalgebraic number.

The early attempts to determine "the value" of π often led to rational expressions (ratios of integers), not always recognized as merely approximations. It was not until 1761 that π was shown to be irrational by Johann Heinrich Lambert of Alsace (then part of Switzerland). The result did not settle the classical problem of squaring the circle, as some

Table 12.3 Computations of π by Electronic Computers

Date	Programmers	Computer	No. of decimal places
1949	G. W. Reitwiesner, et al. (U.S.)	ENIAC	2,035
1955	S. C. Nicholson and J. Jeenel (U.S.)	NORC	3,089
1958	G. E. Felton (England)	Pegasus	10,020
1959	François Genuys (France)	IBM 704	16,167
1961	J. W. Wrench and Daniel Shanks (U.S.)	IBM 7090	100,265
1966	Jean Guilloud, et al. (France)	STRETCH	250,000
1967	Jean Guilloud, et al. (France)	CDC 6600	500,000
1974	Jean Guilloud, et al. (France)	CDC 7600	1,000,000

irrational numbers (e.g., $\sqrt{2}$) are constructible with ruler and compass. However, every constructible number is also *algebraic*—i.e., a root of a polynomial equation with rational coefficients. More than a century after Lambert, π was finally shown to be nonalgebraic by C. L. F. Lindemann of Munich. Thus a problem that had occupied the attention of professional and amateur mathematicians for more than 2000 years was officially laid to rest. (It still occupies some amateurs who refuse to believe that the problem is unsolvable.)

There is a certain element of fiction in our modern way of talking about "determining the value of π." Many of the ancient writers were really trying to determine the circumference of a particular circle, and even those who were searching for a universal constant did not call it π. In the seventeenth century the English mathematicians William Oughtred, Isaac Barrow, and David Gregory used the symbol π variously for the circumference or half-circumference of a circle (it probably stood for "periphery"), but the first to use it for the ratio of circumference to diameter was another Englishman, William Jones, in 1706. The notation did not really catch on, however, until it was adopted by Euler in 1737 and later made popular throughout Europe by his many textbooks.

References

Ball (Chap. 12); Beckmann; Breusch; Coolidge (1953); Eves (1962; 1983, Chap. 4); Gardner (1960a); Halsted; Hogben (Chap. 6); Jones (1950; 1957); Miel; Neugebauer (Chap. 2); Phillips; Read (1967); Schenkman; Schepler; Schoy; Smith, D. E.; Struik; von Baravelle (1952); Wrench (1960).

CHAPTER 13

HOW DID ARCHIMEDES DO IT?

In this chapter we take a close look at the method for computing π that was invented by Archimedes in the third century B.C. and which remained essentially the only successful method for about 1900 years.

For convenience, we consider a circle of diameter 1 (hence of circumference π), and in it we inscribe a regular hexagon. The hexagon is chosen for a starting point because each side of it has length ½. (Why?) Therefore its perimeter is 3 (see Fig. 13.1, and ignore the outer polygon for the time being). At each step in the process of approximating the circle by polygons, we double the number of sides at the previous step. This allows us to use all the same vertices over again and add new ones halfway along the arc of the circle between each pair of old ones (see Fig. 13.2). After k such doubling steps, we have $n = 2^k \cdot 6$ sides. Let a_k denote the sidelength for this number of sides; then the perimeter p_k of the polygon formed at this stage is na_k. Our first task is to determine how to compute a_{k+1} from a_k, for then we can obviously compute as many terms of this sequence as desired, and hence also the sequence of perimeters approximating π.

Exercise

1. Show that

$$a_{k+1}^2 = \frac{1 - \sqrt{1 - a_k^2}}{2}. \tag{13.1}$$

HINT: Apply the Pythagorean Theorem to appropriate right triangles in Figure 13.2.

Starting with $a_0 = 0.5$, and $p_0 = 3$, we can obviously compute terms of the sequence $a_1, a_2, \ldots$ from formula (13.1), after taking square roots on both sides, and thereby also compute $p_1, p_2, \ldots$ from

$$p_k = n_k a_k, \tag{13.2}$$

where

$$n_k = 2^k \cdot 6 = 2n_{k-1}. \tag{13.3}$$

The sequence $p_0, p_1, p_2, \ldots$ is clearly increasing. (Why?) It is also bounded above by π, but that is not so easy to prove directly. Furthermore, it is not clear at this stage how to decide when a given p_k is a satisfactory approximation to the limit.

These difficulties may be overcome by also considering circumscribed polygons with the

Figure 13.1

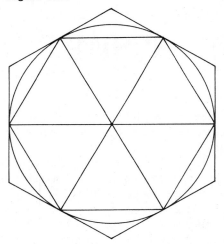

same number of sides as the inscribed ones. Let b_k be the sidelength of a circumscribed regular polygon with n_k sides. If the sides are chosen tangent to the circle at the vertices of the corresponding inscribed polygon (see Fig. 13.1 again), then it is not difficult to determine b_k in terms of a_k.

Exercise

2. Show that

$$b_k = \frac{a_k}{\sqrt{1 - a_k^2}} \,.$$

(13.4)

HINT: See Figure 13.3. Use similar triangles to relate b_k to a_k.

With little extra effort, while we are computing a_k and p_k, we can also compute b_k and the kth circumscribed perimeter $q_k = n_k b_k$. Now it is evident that the sequence $q_1, q_2, q_3, \ldots$ is decreasing. (Why?) It would also appear that

$$p_k < \pi < q_k$$

(13.5)

for each k. (Note that a side of the inscribed polygon is shorter than the subtended arc of the circle because a straight line is the shortest distance between two points. One would think that a side of the circumscribed polygon would be *longer* than the arc it subtends, but the same argument does not apply. You will find a formal proof of this fact, perhaps slightly disguised, in the section of your calculus book in which the formula for differentiating the

Figure 13.2

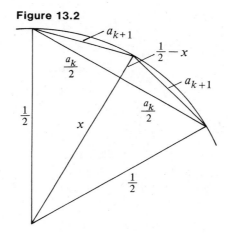

sine function is derived.) Hence, if $q_k - p_k$ is less than some specified tolerance ϵ, it follows that both p_k and q_k are within ϵ of π. It is not so easy to show directly that $q_k - p_k$ is small for k sufficiently large, even though this is intuitively clear. However, we don't really need to: The computer can tell us when $q_k - p_k$ is less than our specified ϵ, and then both q_k and p_k will be within ϵ of π. (There is, however, an important technical problem to be noted here. A computer can carry only a certain fixed number of significant digits in all its computations, and this number depends on the particular computer system being used. Thus every step of a computation is subject to a small truncation or roundoff error. The perimeter method for computing π is unfortunately subject to *accumulation* of these errors, which can make it impossible for the computer to decide accurately whether $q_k - p_k$ is, in fact, less than ϵ, particularly if ϵ is very small or k is very large. This problem is discussed in more detail in the next chapter.)

Figure 13.3

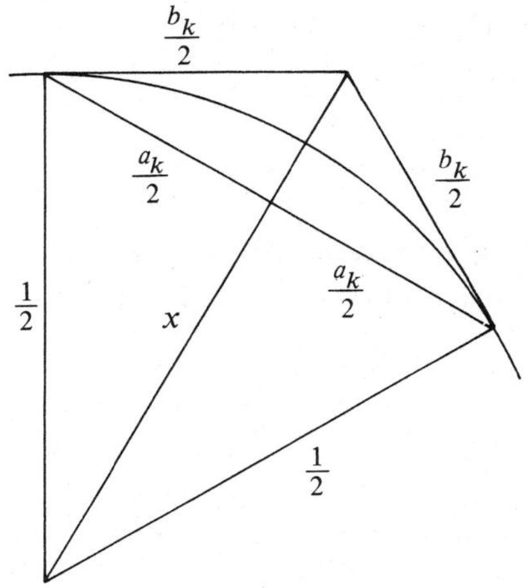

Figure 13.4

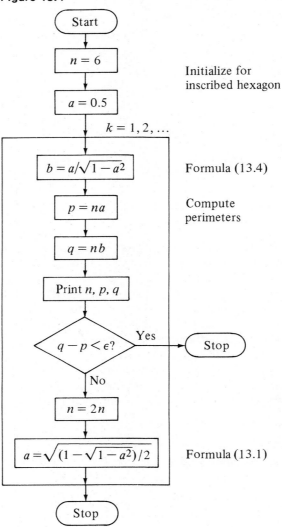

Start

$n = 6$ — Initialize for inscribed hexagon

$a = 0.5$

$k = 1, 2, \ldots$

$b = a/\sqrt{1 - a^2}$ — Formula (13.4)

$p = na$ — Compute perimeters

$q = nb$

Print n, p, q

$q - p < \epsilon?$ — Yes → Stop

No

$n = 2n$

$a = \sqrt{(1 - \sqrt{1 - a^2})/2}$ — Formula (13.1)

Stop

The algorithm represented by formulas (13.1) through (13.4) and the preceding discussion is summarized in the flowchart shown in Figure 13.4.

Exercises

3. Write a program to implement the flowchart of Figure 13.4. If your computer system uses single-precision arithmetic (as most do), take $\epsilon = 0.0005$. However, if you have access to double- (or multiple-, or extended-) precision arithmetic (check with your instructor), you can use a much smaller ϵ, say, 10^{-8}.

4. [O] One might suppose that if p and q are close to π, and π lies between them, then the average $(p + q)/2$ will be even closer to π. Add this quantity to the output line in your program.

References

Miel; Phillips; Polya; Schreiner (Chap. 3); Te Selle.

CHAPTER 14

A COMPUTATIONAL PIECE OF PI [L]

The program you wrote for Exercise 3 in Chapter 13 allows you to repeat the experience of Archimedes in computing π, but of course in much less time than it took him. In fact, with a small modification of the program (see following discussion), you can match what Archimedes' successors accomplished in hundreds of years (with a little help from your friend).

Type in your program and execute it. Do your answers for p and q agree with π to an appropriate number of places? (A correct value to many more places was given at the beginning of Chap. 12 for reference.) How many additional correct places do you get by averaging p and q?

Your curiosity may well have been aroused by our insistence on rather modest choices for ϵ and by the distinction between single and double precision. The problem here is that, for large values of k, p is being computed as the product of a very large number (n) and a very small number (the sidelength, a). Furthermore, each new a is being computed from an old one (13.1) by taking the square root of a much smaller quantity. There will be some roundoff error in that smaller quantity that will be magnified by the square root step, and then further magnified by the multiplication by n.

The choices we have proposed for ϵ cause the program to stop short of the point at which these errors would affect the computed answers. The reason that double precision permits a smaller ϵ, and therefore a more accurate computation of π, is that twice as many significant digits are being carried in all intermediate computations, and so the troublesome roundoff error takes place in a more distant decimal place.

If you really *must* see how roundoff error can overwhelm the algorithm and spoil the results, replace the ϵ in your program by 10^{-4} times itself (i.e., move the decimal points four places to the left). Then execute the program again. Check carefully to see how many correct places there are in each answer. Do the p's and q's still obey the inequality (13.5) of Chapter 13?

Now let's see how we can overcome the magnified error problem, as previously described, *without* resorting to double-precision arithmetic. The appropriate tool is a simple algebraic manipulation called "rationalizing the numerator." You may have learned in high school that one should always rationalize the *denominator*. Why? Because the answer was "wrong" (or at least not quite right) if you didn't do it. Actually, in a precalculator age there was some justification for preferring "$\sqrt{2}/2$" to "$1/\sqrt{2}$": If you already know that $\sqrt{2}$ is approximately 1.414, it is easy to tell by inspection that the former is approximately 0.707, whereas the latter expression would require a long division to determine a decimal approximation. But with a calculator in hand, that justification evaporates. On the other hand, your calculus course provides many occasions for rationalizing a numerator, and this digression has been an introduction to such an instance.

Look again at formula (13.1). When a_k is small, a_k^2 is very small, so $1 - a_k^2$ is close to 1, and its square root is even closer to 1. Thus the subtraction in the numerator results in a loss of significant digits. For example, if the computer works with eight decimal digits in each number, then

$$1.0000000 - 0.99999672 = 0.00000328.$$

The number on the right has only three significant digits, even if the computer records it as

$$0.32800000 \times 10^{-5}.$$

The error in the fourth and subsequent digits is the error we referred to previously that gets magnified by the square root step and then again by the multiplication step.

Now rationalize the numerator in formula (13.1) by multiplying and dividing by $1 + \sqrt{1 - a_k^2}$. The result is

$$a_{k+1}^2 = \frac{a_k^2}{2 + 2\sqrt{1 - a_k^2}}. \tag{14.1}$$

(Verify.) Notice that this formula has *no* subtraction of nearly equal numbers. Even though it computes exactly the same thing as formula (13.1) in theory, its numerical performance can be expected to be quite different. Try it now: Replace the line in your program that updates A by the update implied by equation (14.1):

$$\text{LET A} = \text{A/SQR}(2 + 2 * \text{SQR}(1 - \text{A} * \text{A}))$$

Set ϵ to 5×10^{-8} and run again. Now how many correct places do you observe in P and Q?

[O] The new, more accurate version of the algorithm can also be improved by computing in double precision. If you have access to double-precision arithmetic, try the program again to see how many correct digits you can compute.

WARNING: Some microcomputers offer double-precision variables but only single-precision functions. It is very important that the square roots also be computed in double precision.

SUGGESTION: Turn back to Section 4.4, which presents the "Babylonian" square root method. SQR(X) can be taken as an accurate starting "guess" for the method, which can then be carried out for a fixed (small) number of steps with double-precision variables only, to get a double-precision square root of X. It helps if you know about (or learn about) subroutines for this task, but the program can be written without a subroutine.

References

Forsythe; Franta; Hamming.

CHAPTER 15

DIGGING FOR ROOTS

One of the most important problems of mathematics throughout the ages has been the problem of "solving equations," i.e., of finding the values of one or more "unknowns" that turn an equation into an identity. A special case of this problem is the following. Given a function $f(x)$, find the real numbers r (if any) such that $f(r) = 0$. Such a number r is called a *zero* of f or a *root* of the equation $f(x) = 0$. More generally, any equation in one unknown (variable) has the form $f(x) = g(x)$, where f and g are functions, and r is a solution of this equation if and only if it is a root of $F(x) = 0$, where $F(x) = f(x) - g(x)$. So finding zeros of functions is an equivalent problem to finding solutions of equations in one variable.

You may have already encountered this problem in your calculus course, in connection with the problem of maximizing or minimizing a function. The calculus method for doing that involves computing a derivative f' and then solving the equation $f'(x) = 0$. Of course, all such problems in your calculus text are carefully constructed so that the roots of f' are either obvious or easily found (by factoring, say). But the real world doesn't work that way; in general, you cannot use the calculus max/min technique unless you can also find roots, at least approximately.

In this chapter we describe a very crude method for the approximate determination of a root. The method would be rather useless for all practical purposes if it were not for the fact that we have a computer available to do the arithmetic, because the sequences of numbers generated converge rather slowly to the desired roots. In Chapter 18 we see how the use of differential calculus adds an important new (but 300-year-old) dimension to quick and accurate determination of roots.

One might suppose that if f is a "reasonable" function (continuous, say), and we know that $f(a) < 0$ and $f(b) > 0$ (where $a < b$), then there must be some number r between a and b

Figure 15.1

(b, f(b))
a
r
b
(a, f(a))
(b, f(b))
a
c
b
(c, f(c))
(a, f(a))

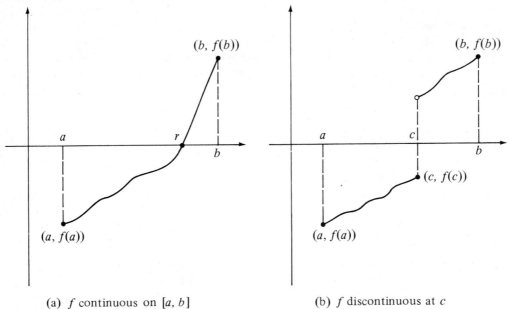

(a) f continuous on $[a, b]$ (b) f discontinuous at c

Figure 15.2

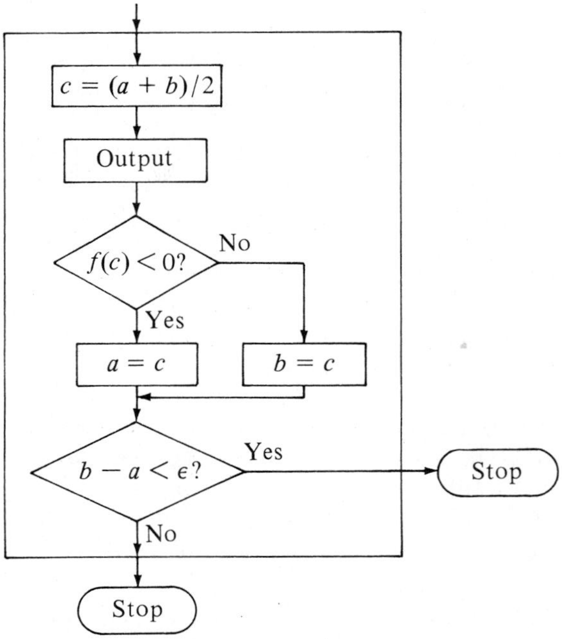

c = (a + b)/2
Output
f(c) < 0?
No
Yes
a = c
b = c
b − a < ε?
Yes
Stop
No
Stop

where $f(r) = 0$. Figure 15.1 illustrates the role that continuity plays in this supposition. Actually, this supposition is correct—indeed, it is one version of an important fact about real numbers called the Intermediate Value Theorem (about which we have more to say in Chap. 17).

Suppose we have the situation of Figure 15.1(a): a continuous function f on $[a, b]$ with $f(a) < 0 < f(b)$ and an unknown root r. In order to "know" r, it is sufficient to know a very small interval $[a', b']$ containing r. For example, if $(b' - a') < 0.00005$ and $a' < r < b'$, then the first four decimal digits of r agree with those of a' (or b'). Our original interval $[a, b]$, obtained perhaps by guessing, by trial and error, or by graphing f, will probably not be very small. However, if we set $c = (a + b)/2$, then the intervals $[a, c]$ and $[c, b]$ are half as long as $[a, b]$, and one of them contains r. (Why?) How can we tell which one? Because we know f, we can compute $f(c)$. If $f(c) < 0$, then r is in $[c, b]$, and if $f(c) > 0$, then r is in $[a, c]$. (Why?) Whichever is the case, we now know an interval of half the length of $[a, b]$ that contains r. Now repeat the process as many times as necessary. This produces a sequence of intervals of lengths $(b - a)/2$, $(b - a)/4$, $(b - a)/8$, . . . , each containing r. Eventually, we must find one of a length less than 0.00005, e.g., because $(b - a)/2^n \rightarrow 0$ as $n \rightarrow \infty$.

The essential part of the interval-halving algorithm for finding a root is summarized in the partial flowchart shown in Figure 15.2.

Exercises

1. Complete the flowchart of Figure 15.2 (definition of f, input step for a and b, etc.), and write a program to implement it. The output should include at least c (the current best estimate of r) and $f(c)$ (which should eventually get close to 0). Take $\epsilon = 0.00005$.

2. [O] In the discussion thus far, we have, in the interest of simplicity, deliberately ignored some minor details. One of these is what to do if the value of f at a or b or c is actually zero—i.e., the true root has accidentally been found. This is unlikely to happen unless you deliberately choose a problem with a rational solution that you know in advance and carefully choose the initial a and b. However, you may want to modify your program slightly to take advantage of serendipity. Using a picture similar to Figure 15.1(a), you might consider what could happen if f at an endpoint turns out to be 0 and you ignore it.

3. If you have followed each step carefully this far—and if you have not been thinking ahead—you now have a root-finding program that assumes a continuous function f on an interval $[a, b]$ with $f(a) < 0 < f(b)$. Of course, we also want to be able to handle the case $f(a) > 0 > f(b)$. Decide now what you want to do about this. [One possibility: Replace f by $-f$ and use your present program without change. Another: Check carefully to see what your program does if a is the *right-hand* endpoint and b the *left-hand* one. Still another: Base your decision-making test on the sign of $f(a)f(c)$ rather than just $f(c)$—i.e., consider whether $f(c)$ has the same sign as $f(a)$.]

4. [O] A variation of interval-halving of interest to anyone who works primarily with *decimal* arithmetic (i.e., everyone) is obtained by dividing the interval $[a, b]$ into *ten* equal subintervals, rather than two, and finding the subinterval on which a sign change occurs. Note that if the original interval has length 1, each subdivision step adds one more decimal digit to the final answer. (Why?) Make a flowchart for, and program, this variation.

C H A P T E R 1 6

LET'S COMPUTE ROOTS [L]

This laboratory session consists of several root-finding problems to be solved by using your program from Chapter 15. Each problem may require some advance preparation, such as a sketch of the particular function, to help you determine starting values for a and b. The last two exercises also require prior determination of the function.

If you decide to use the GRAPH program from Chapter 6 to help you select a and b, remember to save your root-finding program before loading GRAPH.

Exercises

1. Find $\sqrt{2}$ by solving $x^2 - 2 = 0$. (This is a test problem to make sure your program works as it should.)

2. Find all three roots of

$$x^3 - 10x^2 + 22x + 6 = 0.$$

3. Find a root of

$$x = \cos x$$

between 0 and 2. [Set $f(x) = x - \cos x$. How do you know that $f(0) < 0 < f(2)$?]

4. The polynomial equation

$$179.53x^6 - 436.85x^4 + 257.32x^2 - 0.001 = 0$$

has six real roots, all between -2 and 2. How many can you find?
HINT: $f(x) = f(-x)$.

5. Find the smallest positive root of the equation

$$x = \tan x.$$

SUGGESTION: Proceed as in Exercise 3: $y = x$ is tangent to $y = \tan x$ at $(0, 0)$, and the next crossing point of the two graphs occurs between $\pi/2$ and $3\pi/2$. Those numbers, however, will not do for a and b. (Why?) (Your system may or may not have a built-in tangent function. If not, it may be programmed as SIN(X)/COS(X).)

6. Find a root of

$$2 + \cos x = (x + 1)^3$$

between 0 and 1.

7. Find a root of

$$\tan x = 2 - x^2$$

between 0 and 1.

8. Find the minimum value of the function

$$f(x) = \frac{10}{x^2} + 3x^2 + x, \qquad x > 0.$$

NOTE: This problem looks as if it might have come from the max/min section of any calculus book, but it didn't because it is not obvious how to find a root of the derivative f'.[1]

9. The following problem looks like one of those silly word problems from typical high school algebra texts. However, the equation one finds for the unknown turns out to be a cubic that is not amenable to solution by the usual elementary means.

> A flag pole stands one hundred feet high and ten feet from a ten foot pole. If the flag pole breaks over (top section does not separate from base section) in such a way that it touches the top of the ten foot pole and just touches the ground, find the height of the break.[2]

---

[1] It is, however, used as an example in a section on numerical determination of roots in S. Bell, J.R. Blum, J.V. Lewis, and J. Rosenblatt, *Modern University Calculus,* Holden-Day, San Francisco, 1966, pp. 393–394.

[2] The problem is quoted from B.E. Meserve, *Fundamental Concepts of Algebra,* Addison-Wesley Publishing Company, Inc., Reading, MA, 1953, p. 171.

C H A P T E R   1 7

THE INTERMEDIATE VALUE THEOREM [0]

Let's take a closer (mathematical) look at the interval-halving algorithm of Chapter 15. We are considering a continuous function f on a closed interval $[a, b]$, with $f(a) < 0 < f(b)$. The algorithm constructs a sequence of intervals $[a_1, b_1]$, $[a_2, b_2]$, ..., or, equivalently, *two* sequences of numbers

$$a_0, a_1, a_2, \ldots$$

and

$$b_0, b_1, b_2, \ldots,$$

where $a_0 = a$, $b_0 = b$, and in general, a_{n+1} and b_{n+1} are defined by the following conditions:

1. If $f((a_n + b_n)/2) < 0$, then $a_{n+1} = (a_n + b_n)/2$, and $b_{n+1} = b_n$.

2. Otherwise (i.e., if $f((a_n + b_n)/2) \geq 0$), then $a_{n+1} = a_n$, and $b_{n+1} = (a_n + b_n)/2$.

(Fig. 17.1 illustrates the first four steps of this process.)

Exercises

1. Show that both of the sequences $\{a_n\}$ and $\{b_n\}$ are monotone and bounded and hence convergent (see Property 9.6).

2. Show that $\lim_{n \to \infty} a_n = \lim_{n \to \infty} b_n$ (see Property 9.5 and Example 7.12).

3. Show that $\lim_{n \to \infty} f(a_n) = \lim_{n \to \infty} f(b_n)$ (see Theorem 11.3).

4. If $L = \lim_{n \to \infty} a_n$, show that $f(L)$ is ≤ 0. Similarly, show that $f(L) \geq 0$. Conclude that $f(L) = 0$ — i.e., L is a root of $f(x) = 0$.

A quick glance backward reveals that the last sentence of Exercise 4 contains the first mention of a *root* in this chapter. Rather than assume the existence of such a root, we have used the algorithm to prove that the root exists by constructing it as the common limit of the two sequences. We summarize the development in the exercises with the following version of the *Intermediate Value Theorem:*

If f is a continuous function on $[a, b]$, and $f(a) < 0 < f(b)$, then f has a root in $[a, b]$.

Figure 17.1

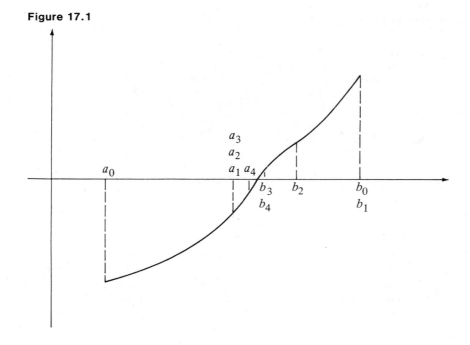

Of course, the same conclusion can be drawn if $f(a) > 0 > f(b)$ by applying the previous result to $-f$. The usual version of the Intermediate Value Theorem is the following (apparently) more general statement.

Intermediate Value Theorem   If f is continuous on $[a, b]$, and d is any number between $f(a)$ and $f(b)$, then there is some number c in $[a, b]$ such that $f(c) = d$.

Proof   Apply the previous result to the function $g(x) = f(x) - d$, to conclude that g has a root c in $[a, b]$. Then $g(c) = f(c) - d = 0$, so $f(c) = d$.

Remark   This development of the Intermediate Value Theorem leans heavily on Property 9.6, the convergence of bounded monotone sequences. Please keep in mind that this property has been *assumed* as the subtle completeness property of the real numbers.

Reference

Grabiner (especially Sect. 4).

C H A P T E R   1 8

UNDER THE APPLE TREE, OR HOW ISAAC NEWTON LOOKED FOR ROOTS

18.0 HISTORICAL BACKGROUND [R]

In Chapter 15 we took our first look at the problem of finding roots, i.e., solving equations of the form $f(x) = 0$. The method of interval-halving rests on the understanding of continuous functions and of limits developed in the nineteenth century; but as a practical computational device, it is obviously much more recent, as the use of a computer is almost essential. Prior to the invention of digital computers, many efforts were made over a period of 4000 years to develop methods for the practical solution of equations. The results of these efforts are still of interest; and that we have a powerful computational tool available is no excuse for using it inefficiently.

One line of development in the historical search for roots has been the attempt to find explicit formulas for solving equations, especially polynomial equations. One reason for the development of a square root method in ancient Mesopotamia[1] was that the Babylonians knew how to solve quadratic equations by using essentially the same quadratic formula we use today.[2] In the third century B.C., Archimedes gave a complete analysis of cubic equations of the form $x^2(c - x) = a$, which arose in connection with a geometric problem. In a sense, this amounted to a complete analysis of cubic equations, as every cubic can be reduced to one of the indicated form. However, the first explicit solution of the general cubic equation was the result of some remarkable algebraic activity in sixteenth-century Italy.

In 1545 Geronimo Cardano published the *Ars magna,* which contained the complete solutions of the general cubic and the general quartic as well! Neither solution was Cardano's own work. The cubic formula he attributed to Niccolo Tartaglia who discovered it in 1541; but Cardano failed to mention that he had promised not to publish it before Tartaglia had a chance to do so himself. It is now known that a general cubic formula was known earlier to Scipione del Ferro, who died in 1526. The quartic formula Cardano attributed to his employee Ludovico Ferrari, "who invented it at [Cardano's] request."[3]

The simultaneous publication of complete solutions for cubics and quartics sparked

---

[1]See Section 4.4.

[2]The knowledge of this method in the Western world, however, is due essentially to the influence of al-Khowarizmi's book on algebra (see Chap. 3).

[3]These formulas are too complicated to reproduce here. General methods for solving cubics and quartics may be found in Meserve (see Bibliography), pp. 150–156. For more details of the controversy surrounding the cubic formula, see Feldmann.

interest in moving on to the general quintic and higher-degree equations, an interest that lasted for more than 250 years. In 1799, another Italian, Paolo Ruffini, published a proof that a general quintic formula is impossible, but the proof was defective. About 20 years later, Niels Henrik Abel, a young Norwegian genius, thought he had found a general quintic formula but later came to the opposite conclusion and, at the age of 19, gave the first correct proof that no general quintic (or higher-order) formula is possible.[4]

The other, more practical, line of development is the search for approximations to roots. As we have noted, the Babylonians knew how to solve quadratic equations and worked out an approximation method for computing square roots in order to complete the job. They also knew how to obtain approximate numerical solutions for certain cubics. In the third century A.D., mathematicians in China achieved approximate solutions of polynomial equations by a method known as *fan fa*. In the West this method is called "Horner's method," after the English mathematician W. G. Horner who published it in 1891, unaware that the method had been used for many centuries in China. By the thirteenth century, *fan fa* was well known among Chinese mathematicians; and in 1303 Chu Shih-chieh found solutions of equations up to the fourteenth degree with "Horner's method."

Perhaps the most significant advance in the search for (approximate) roots was the introduction of calculus techniques by Isaac Newton in 1671. One of Newton's contemporaries, Joseph Raphson (1648?–1715?), published a simplified and improved version of Newton's method in 1690, and it is Raphson's version that is widely known today as "Newton's method," largely due to the influence of nineteenth-century textbook writers who ignored the distinction.[5] Although more properly designated the *Newton-Raphson* method, we use the shorter, more popular name.

This application of the derivative is the subject of this chapter. We see that application of Newton's method to equations of the form $x^2 - a = 0$ leads to the ancient Babylonian square root method. For polynomial equations in general, it is essentially equivalent to the Chinese "Horner" method. But one of the significant advantages of the method is that it applies to most equations of the form $f(x) = 0$ with f differentiable, regardless of whether or not f is a polynomial. The method is less general than interval-halving, in that differentiability is required; but this is more than offset by a much more rapid rate of convergence, as we see empirically in the next laboratory session and formally in Chapter 20. There is, however, a price to be paid for rapid convergence: Unlike interval-halving, the method does not always converge. Hence we must confront a new kind of problem, namely, the determination of conditions that will assure convergence.

18.1 DESCRIPTION OF THE METHOD

The basic idea behind Newton's method is illustrated by Figure 18.1. Given an initial guess x_0 in the vicinity of an unknown root r of f, we approximate the graph of f by the tangent line

---

[4]About a year earlier, his father died, leaving him the sole supporter of a large, impoverished family. See Prielipp for more about the tragic life of Abel who died at the age of 26 and left many important contributions to several branches of mathematics. Unfortunately, these were mostly unrecognized in his lifetime.

[5]See Cajori (1911) for details of what Newton actually did and how Raphson modified it.

Figure 18.1

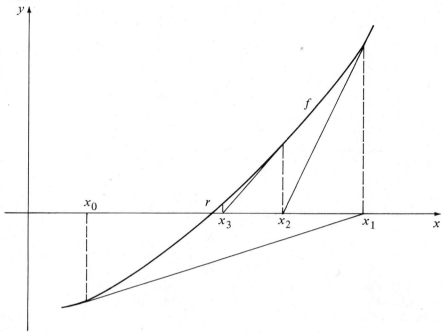

at $(x_0, f(x_0))$:

$$y - f(x_0) = f'(x_0)(x - x_0). \tag{18.1}$$

The next approximation x_1 to r is the point where the tangent line crosses the x-axis. If we substitute $x = x_1$, $y = 0$ in formula (18.1) and solve for x_1, we find

$$x_1 = x_0 - \frac{f(x_0)}{f'(x_0)}, \tag{18.2}$$

provided, of course, $f'(x_0) \neq 0$. (What happens if $f'(x_0) = 0$?)

Now the process repeats: Starting with x_1, x_2 is defined as the point where the tangent line at $(x_1, f(x_1))$ crosses the x-axis. The computation is the same as that leading to formula (18.2), with each subscript increased by one. In general, the $(k + 1)$-th approximation to r is computed from the kth by

$$x_{k+1} = x_k - \frac{f(x_k)}{f'(x_k)}. \tag{18.3}$$

Of course, we need to ensure that $f'(x_k) \neq 0$ for any k.

Figures 18.2 and 18.3 illustrate two of the ways the sequence $\{x_k\}$ can fail to converge to r. In Figure 18.2 the sequence oscillates between two numbers, essentially because $f'(x)$ takes the value 0 between x_0 and r, even though $f'(x_k)$ is never 0. The same sort of oscillation occurs in Figure 18.3, even though $f'(x) \neq 0$ anywhere. This time the problem is caused by

Figure 18.2 **Figure 18.3**

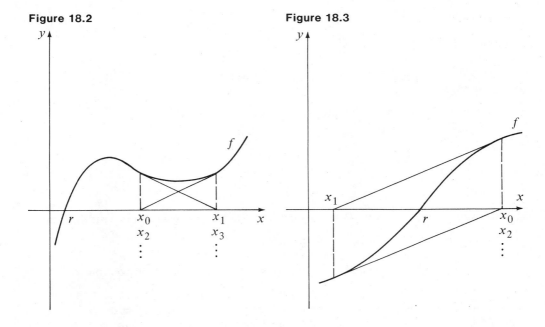

the presence of an inflection point. These are not the only ways in which the method can fail to converge; obviously we need to find out how to distinguish "good" situations in which to use it (e.g., Fig. 18.1) from "bad" ones.

Before tackling the difficult problem of careful analysis of the method, let's solve the easy one: writing a program. Notice that the main step in the algorithm is formula (18.3), which tells how to compute a new x from an old one. Note also that two functions have to be defined, f and f'. (It is obviously much easier for you to compute f' from f than it is to tell the computer how to do it.) We also have to decide when to stop computing new values of x. A reasonable guess when computing a sequence recursively is that, when x_{k+1} is very close to x_k, continued repetitions will not change things very much, i.e., the limit must be nearby. This is sometimes correct—e.g., if the limit is always between x_k and x_{k+1}—and sometimes is not correct; so our analysis of the method must also take up the validity of the stopping rule.

Exercise

1. Make a flowchart for Newton's method. Take into consideration appropriate function definitions, input of the initial guess x_0, and the stopping tolerance ϵ. Put an output step inside the loop so that the rate of convergence can be checked. Exit from the loop when $|x_{k+1} - x_k| < \epsilon$. Now convert the flowchart into a program.

The next section (designated optional, for those who need to be convinced) establishes the truth of the following theorems. (The proofs require some facts from Chaps. 9 and 11.)

Theorem 1 Suppose there is some interval I on the real line such that

1. $f''(x)$ exists for all x in I.

2. $f'(x) \neq 0$ for all x in I.

3. x_0 and r are in I.

4. There is some constant $M < 1$ such that

$$\left| \frac{f(x) f''(x)}{f'(x)^2} \right| \leq M$$

for all x in I.
Then $r = \lim_{k \to \infty} x_k$.

Intuitively, what condition (4) says is this: To ensure convergence, it is desirable that

1. x_0 be close to r, so that $|f(x)|$ stays small.
2. $|f'(x)|$ be large near r; i.e., steep slopes are preferred.
3. $|f''(x)|$ should stay small; i.e., the slope should not be changing too rapidly.

Unfortunately, condition (4) gives no insight into why an inflection point near the root can cause trouble, as illustrated in Figure 18.3. Suffice it to say, the difficulty posed by that example is an optical illusion. It *is* possible for a bad choice of x_0 to lead to such an oscillation, but there is *no* differentiable function like the one shown in Figure 18.3, for which *every* choice of x_0 leads to oscillation.[6] Thus the difficulty can be made to disappear by a better choice of a starting point.

Theorem 2 Suppose the conditions of Theorem 1 are all satisfied, and moreover $M \leq \frac{1}{2}$. Then, whenever $|x_{k+1} - x_k| < \epsilon$, we also have $|x_{k+1} - r| < \epsilon$.

In other words, under certain conditions there is no doubt that the end test in the program will ensure that we find the desired root to the desired accuracy. Unfortunately, it is often very difficult to check that these conditions are actually satisfied.

Example (Square roots) Suppose $f(x) = x^2 - a$ with $a > 0$. Then $f'(x) = 2x$ and $f''(x) = 2$. We have

$$x_{k+1} = x_k - \frac{f(x_k)}{f'(x_k)} = x_k - \frac{x_k^2 - a}{2x_k} = \frac{1}{2}\left(x_k + \frac{a}{x_k}\right).$$

This is precisely the recursion for the Babylonian square root method (see Sect. 4.4). We also have

$$\left| \frac{f(x) f''(x)}{f'(x)^2} \right| = \left| \frac{2(x^2 - a)}{4x^2} \right|$$

$$= \frac{1}{2}\left| 1 - \frac{a}{x^2} \right|.$$

---

[6]See Weiner.

If $x \geq \sqrt{a}$, $0 \leq |1 - a/x^2| < 1$, so

$$\left| \frac{f(x) f''(x)}{f'(x)^2} \right| \leq \frac{1}{2}$$

on the interval $[\sqrt{a}, \infty)$. This shows that any starting value $x_0 > \sqrt{a}$ leads to convergence. However, if $0 < x_0 < \sqrt{a}$, then $x_1 > \sqrt{a}$, so *any* positive starting guess leads to convergence (see Sect. 11.4, Exercise 9). This is an alternative proof to the one sketched in (optional) Case Study 11.4.

18.2 DOES IT WORK? [O]

Before you take your program to the computer, let's see what we can do to create some confidence that the method actually works and to counter the doubts deliberately planted by Figures 18.2 and 18.3.

As stated in Theorem 1, and for reasons that will become clear, let's suppose that f is *twice* differentiable on some interval containing both x_0 and r and that $f'(x) \neq 0$ on the interval. To determine whether the sequence $\{x_k\}$ converges to r, we want to compare $|x_{k+1} - r|$ to $|x_k - r|$. The tool for doing this is the Mean Value Theorem.

Define a new function g on the interval in question by

$$g(x) = x - \frac{f(x)}{f'(x)}.$$

Then formula (18.3) can be rewritten as

$$x_{k+1} = g(x_k). \tag{18.4}$$

Furthermore, because $f(r) = 0$, we have

$$g(r) = r. \tag{18.5}$$

Exercise

2. Show that $g'(x) = f(x) f''(x) / f'(x)^2$.

Because g is a differentiable function, the Mean Value Theorem applies to it, and we have

$$
\begin{aligned}
|x_{k+1} - r| &= |g(x_k) - g(r)| &&\text{[by equations (18.4) and (18.5)]} \\
&= |g'(c_k)(x_k - r)|, &&\text{for some } c_k \text{ between } x_k \text{ and } r, \\
&= |g'(c_k)| \, |x_k - r|.
\end{aligned}
$$

Now let's suppose we can find some number $M < 1$ so that $|g'(x)| \leq M$ for all x in a suitably small interval containing x_0 and r. Then the preceding computation shows that

$$|x_{k+1} - r| \leq M |x_k - r|, \qquad \text{for each } k,$$

and hence,

$$|x_1 - r| \le M|x_0 - r|,$$
$$|x_2 - r| \le M|x_1 - r| \le M^2|x_0 - r|,$$
$$|x_3 - r| \le M|x_2 - r| \le M^3|x_0 - r|,$$

and so on. In general,

$$|x_k - r| \le M^k|x_0 - r|.$$

Because we required M to be less than 1, $\lim_{k \to \infty} M^k = 0$, by Theorem 11.1. It follows that $\lim_{k \to \infty}|x_k - r| = 0$, and hence that $\lim_{k \to \infty} x_k = r$ (Properties 9.7 and 9.3). This completes the proof of Theorem 1.

Now let's see what we can say about our end test: Does $|x_{k+1} - x_k| < \epsilon$ ensure that $|x_{k+1} - r|$ is also small, as indicated in Theorem 2? If M is selected as before, we have

$$
\begin{aligned}
|x_{k+1} - r| &\le M|x_k - r| \\
&\le M(|x_k - x_{k+1}| + |x_{k+1} - r|) \qquad \text{(``Triangle Inequality'')} \\
&\le M\epsilon + M|x_{k+1} - r|.
\end{aligned}
$$

Hence

$$(1 - M)|x_{k+1} - r| < M\epsilon,$$

and because $M < 1$,

$$|x_{k+1} - r| < \frac{M\epsilon}{1 - M}. \tag{18.6}$$

For M close to 1, formula (18.6) doesn't say much. But if $M \le \frac{1}{2}$, then $M/(1 - M) \le 1$ (check it!), and then formula (18.6) tells us that our last computed x is within ϵ of r. This completes the proof of Theorem 2.

Unfortunately, it is usually not so easy to determine a suitable interval and a "magic number" M bounding $|g'(x)|$ as it is for the square root case (Sect. 18.1, Example). Indeed, this task is sufficiently difficult in general that it makes more sense to try the computer first, interrupt it if convergence fails to take place, and then try to determine what caused the failure. The formal analysis should be taken as evidence that there *are* reasonable conditions under which the method is sure to work. A rough sketch of the function f whose roots are to be found is always helpful for suggesting starting points and also tipping us off as to possible difficulties with either the first or second derivative.

18.3 CUBE ROOTS [O]

The following exercises guide you through the application of Newton's method to the computation of cube roots. The steps are analogous to those of the square root example in Section 18.1, and they further illustrate the theoretical development.

Exercises

3. Derive the following recursion formula for computing cube roots of positive numbers a, by applying Newton's method to the function $f(x) = x^3 - a$:

$$x_{k+1} = \frac{2}{3} x_k + \frac{a}{3x_k^2}.$$

(Note that this is a "weighted average" of x_k and a/x_k^2.)

4. Show that

$$\left| \frac{f(x) f''(x)}{f'(x)^2} \right| = \frac{2}{3} \left| 1 - \frac{a}{x^3} \right|.$$

Conclude that $M = \frac{2}{3}$ is a suitable "magic number" on the interval $[\sqrt[3]{a}, \infty)$.

5. Show that, if $| x_{k+1} - x_k | < \epsilon$, then $|x_{k+1} - \sqrt[3]{a}| < 2\epsilon$.

6. Write a program to compute cube roots, using the recursion of Exercise 3.

References

Cajori (1911); Coate; Feldmann; Henriksen; Kuller; Mott; Parker; Prielipp; Struik; Traub; Weiner.

CHAPTER 19

SOLVING EQUATIONS, OLD AND NEW [L]

The exercises in this laboratory session are root-finding problems to be solved by Newton's method by using your program from Exercise 18.1. Some advance preparation is required. In particular, in each case you have to compute a derivative of the appropriate function. You also have to decide on a starting value of x and an appropriate tolerance ϵ for each problem. In some cases even a very crude guess for x works, but in others you may have to sketch a rough graph of the function to get some idea of a starting value. (Alternatively, use the GRAPH program for this task.) Your choice of ϵ should permit five or six significant digits in your answer, but what this means depends on where you expect to find a root (e.g., there is a big difference between looking for a root between 0 and 1 and looking for one near $x = 100$). If your system works in double precision, much better accuracy is possible. (But who needs it?)

Several of the exercises are repeated from Chapter 16 to allow you to compare Newton's method with interval-halving. In particular, two of these involve trigonometric functions. If you have not yet studied differentiation of trigonometric functions, you can look up the appropriate differentiation formulas in your calculus book and use them to calculate the necessary derivatives for Newton's method.

Exercises

1. Find $\sqrt{2}$ by solving $x^2 - 2 = 0$ (program checkout problem).

2. Early in the eleventh century A.D., the Persian mathematician al-Biruni showed that the problem of inscribing a regular nine-gon in a circle could be reduced to finding the positive root of the equation $x^3 = 1 + 3x$. He then proceeded to approximate this root to six-decimal-place accuracy. Find the root.

3. In 1225 A.D., Fibonacci (see Sect. 4.3) demonstrated the value of numerical methods to his European contemporaries as follows: He first showed that $x^3 + 2x^2 + 10x = 20$ has no solution of the form $a + \sqrt{b}$ with a and b rational (the only sense in which equations could be solved exactly in his day). He then proceeded to compute a highly accurate approximation to the (unique) root by an unknown method, possibly the equivalent of "Horner's" method. Find the root.

4. About 1600 A.D., François Viète (1540–1603), a French lawyer, politician, and part-time mathematician (but one of the most important of his day), introduced a method for finding roots of polynomials that is similar to, but more cumbersome than, Newton's method. Viète's work was probably Newton's source of inspiration. One of the examples Viète used was the computation of the unique root of $x^5 - 5x^3 + 500x = 7,905,504$. Find the root.

5. Apparently, the only equation that Newton ever bothered to solve with his method (1671) was $x^3 - 2x - 5 = 0$. Find a root near $x = 2$.

6. At Samarkand in the early fifteenth century, the Arab mathematician-astronomer al-Kashi took great delight in showing off his computational skill. One of his feats was to compute the sixth root of the number

$$21,153,274,955,684,299,240\tfrac{2}{3},$$

using "Horner's method," possibly learned from a Chinese source. Find a root to within the accuracy that can be achieved by your computer.

WARNING: Your computer may not accept a number with this many significant digits. If you are allowed only seven significant digits, say, use 2115327E13.

7. Compare Exercise 16.2.
   a. Find a root of $x^3 - 10x^2 + 22x + 6 = 0$, using $x = 1$ as a starting value.

b. Repeat part (a) with a starting value of $x = 2$. Observe that the Newton sequence fails to converge to a root. Work out an explanation (with a picture or with algebra or both) of why convergence fails.

8. Compare Exercise 16.4. Find the three positive roots of

$$179.53x^6 - 436.85x^4 + 257.32x^2 - 0.001 = 0.$$

9. Find two roots of $x^4 - 2x^3 + x^2 - 2 = 0$.

10. Compare Exercise 16.3. Solve $x = \cos x$.

11. Compare Exercise 16.5. Find the smallest positive root of the equation $x = \tan x$.

WARNING: Be careful about selection of a starting point. Consider a graph of $y = x$ and $y = \tan x$ on the same coordinate axes.

12. [O] Let $f(x) = 30x - 1 + (1 - 26x)e^x$. (Skip this problem or come back to it later if you have not studied the exponential function e^x yet.)
   a. Show that $f(0) = 0$ and that the graph of f rises from the origin, then falls to a negative value $f(1)$. Conclude that $f(x) = 0$ has a unique root in the open interval $(0, 1)$.
   b. Find the root.

NOTE: Solution of a problem like this is a necessary step in accurate determination of the time of death of a murder victim by measurements of body temperature when the surrounding temperature has not been constant. See D. A. Smith (1978) for further details.

13. [O] Let k denote the answer to Exercise 12, and let

$$f(t) = (37k - 1 + kt)e^{kt} - 30k + 1.$$

Find a negative root of $f(t) = 0$.

NOTE: The answer to this problem is the time of death referred to in the preceding note measured in hours before discovery of the body. The negative sign means "before." We say more about the context of these two problems in Chapter 49.

14. [O] Find the annual interest rate on a $2000 loan that is to be repaid in 24 monthly payments of $94.15 each.

HINT: Let x be the interest rate as a decimal fraction. Then the monthly rate is $x/12$, and compounding over a two-year period would increase the $2000 borrowed to $2000(1 + x/12)^{24}$ (see Sect. 2.2). However, payments are being made, so the amount borrowed must be reduced by the payments and by the fact that no interest is paid on money already paid back. Thus we subtract $(94.15)(1 + x/12)^{23}$ for the first payment, $(94.15)(1 + x/12)^{22}$ for the second, and so on, down to $(94.15)(1 + x/12)^0$ for the last payment. The final balance is 0, so we have

$$0 = 2000\left(1 + \frac{x}{12}\right)^{24} - (94.15)\sum_{k=0}^{23}\left(1 + \frac{x}{12}\right)^k.$$

Now the sum is a partial sum of a geometric progression, which we can evaluate explicitly, as in Section 11.2:

$$\sum_{k=0}^{23} \left(1 + \frac{x}{12}\right)^k = \frac{1 - (1 + x/12)^{24}}{1 - (1 + x/12)} = -\frac{12}{x}\left[1 - \left(1 + \frac{x}{12}\right)^{24}\right].$$

Substituting in the preceding equation and simplifying, we get

$$\left(2000 - \frac{1129.8}{x}\right)\left(1 + \frac{x}{12}\right)^{24} + \frac{1129.8}{x} = 0.$$

This is the equation to be solved for x. Note that if you first multiply through by x, the equation becomes a polynomial equation, and the appropriate derivative is a little easier to calculate.

15. [O] (For those who completed Exercise 18.6.) Try out your cube root program for various choices of a. Alternatively, you might want to construct a cube root table by enclosing the main part of the program in a loop indexed by a.

16. [O] In Exercises 3 and 4 we asserted that the Fibonacci and Viète equations each have a unique solution. Prove that this is so.

HINT: Use Rolle's Theorem.

Postscript   Did you do Exercise 9? Did you use Newton's method? Why not do it this way?

$$x^4 - 2x^3 + x^2 = 2,$$
$$(x^2 - x)^2 = 2,$$
$$x^2 - x = \pm\sqrt{2}.$$

Hence

$$x^2 - x - \sqrt{2} = 0 \quad \text{or} \quad x^2 - x + \sqrt{2} = 0.$$

One of these two quadratics has no real roots, and the other has the roots

$$\frac{1 \pm \sqrt{1 + 4\sqrt{2}}}{2},$$

which are easily evaluated with a calculator (or a direct inquiry using SQR). It doesn't hurt to think carefully about problems to be solved, and if this were the only problem for which we needed an answer, it would not be worthwhile to program and use Newton's method. (This example was suggested by J. B. Rosser. See Bibliography.)

CHAPTER 20

WHY DOES NEWTON'S METHOD WORK SO WELL? [O]

Our analysis of Newton's method in Chapter 18 was designed to show that, subject to certain reasonable conditions, the method actually works. The analysis was superficially similar to that given earlier (Chap. 15) for the interval-halving method, expressing a bound for the error at the $(k + 1)$-th step as a fractional multiple of the error at the kth step. This gives no clue as to why Newton's method should converge faster than interval-halving, but your experience with the computer should suggest that it is in fact faster.

Actually, our previous analysis of Newton's method was very conservative, mainly because it was based on the Mean Value Theorem. Later in your calculus course you encounter a more general theorem that is variously called Taylor's Theorem or the Extended Mean Value Theorem. The MVT is the "first-order" special case of this result, and the "second-order" case is the following: If f is twice differentiable on $[a, b]$, then

$$f(b) - f(a) = f'(a)(b - a) + \frac{f''(c)}{2}(b - a)^2, \qquad (20.1)$$

where c is some number between a and b. Using this fact, we can get a better estimate of the relationship between $|x_{k+1} - r|$ and $|x_k - r|$, and also better information about the stopping rule.

Exercises

1. Show that

$$|x_{k+1} - r| \le \frac{1}{2}\left|\frac{f''(c_k)}{f'(x_k)}\right| |x_k - r|^2$$

for some c_k between x_k and r. (HINT: Use equation (18.3), and apply (20.1) to $f(r) - f(x_k)$, remembering that $f(r) = 0$.) Conclude that, if N is a number such that $N \ge |f''(t_1)/f'(t_2)|$ for all numbers t_1 and t_2 in some interval containing r, then

$$|x_{k+1} - r| \le \frac{N}{2}|x_k - r|^2. \qquad (20.2)$$

2. Consider the problem of finding a root of $f(x) = x - \cos x = 0$ (Exercise 19.10).
  a. Show that there is a root r in the interval $[0, \pi/2]$. (HINT: Recall the Intermediate Value Theorem.)
  b. Show that $N = 1$ will satisfy the condition stated in Exercise 1 on the interval $[0, \pi/2]$,

and hence

$$|x_{k+1} - r| \leq \frac{1}{2}|x_k - r|^2.$$ (20.3)

c. If x_k agrees with r to two decimal places, how close will x_{k+1} be to r?

3. a. For the function f in Exercise 2, suppose that $|x_{k+1} - x_k| < \epsilon < \frac{1}{2}$, and also that $|x_{k+1} - r| < 1$. Show that

$$|x_{k+1} - r| < \frac{\epsilon^2}{1 - 2\epsilon}.$$ (20.4)

HINT: Start with formula (20.3). Write $x_k - r = (x_k - x_{k+1}) + (x_{k+1} - r)$. Use the Triangle Inequality, multiply out the square, and estimate the terms by using the given inequalities. Note that $|x_{k+1} - r|^2 < |x_{k+1} - r|$. Solve the resulting inequality for $|x_{k+1} - r|$.

b. If the iteration is stopped when $|x_{k+1} - x_k| < 0.0005$, how close is x_{k+1} to r?

C H A P T E R 2 1

CALCULUS FOR FUN AND PROFIT

The following "word problem" has the appearance of a typical exercise from the "applied max/min" section of any standard calculus book.

Problem 1 Suppose that when x hours per day are devoted to a certain manufacturing process, the profit $f(x)$ per dollar invested is given by

$$f(x) = \frac{-26}{x + 2} - \left(\frac{x}{6}\right)^3 + x + 13, \qquad 0 \leq x \leq 24.$$

For example, labor costs might skyrocket if too many hours are put in; other processes might have to shut down, which would cost money; and so on. Find the number x of hours per day that will maximize $f(x)$.

Actually, the problem is taken from a not-so-standard calculus book[1] whose authors suggest the use of a computer in obtaining a solution for reasons that will soon be clear. Let's see what we *can* determine about the function f of Problem 1 by using calculus.

---

[1] S. Bell, et al., *Modern University Calculus*, Holden-Day, Inc., San Francisco, 1966, p. 136.

Exercises

1. Compute $f'(x)$ and $f''(x)$ for the function f defined in Problem 1.

2. Verify each of the following:

        (a) $f(0) = 0$ **(b)** $f(24) < 0$ **(c)** $f'(0) > 0$ **(d)** $f'(24) < 0$

3. Verify that $f''(x) < 0$ for all $x \geq 0$.

4. Make a rough sketch of the graph of f, using the facts verified in Exercises 2 and 3. What can you conclude about critical numbers and possible maxima?

NOTE: The GRAPH program could also be used to obtain this information. In fact, you may have already done so—see Exercise 6.3.

5. Verify that finding critical numbers in this case is equivalent to solving a fourth-degree polynomial equation.

Exercise 5 suggests that one way to solve the problem is to use the root-finding method of Chapter 15 or 18 on f', or on the numerator of f' when expressed as a quotient (cf. Exercise 8 of Chap. 18). This could certainly be done, but one might also ask whether it is not possible to work directly with f to obtain its extreme values.

There is an important reason for asking (and answering) this question: Not every function is differentiable throughout its domain, and extreme values can occur at points where the derivative fails to exist, as well as where $f'(x) = 0$. A method that does not require differentiability (resting, instead, on continuity on a closed interval, say) can handle both kinds of extrema at once.

Suppose we have a continuous function f on a closed interval $[a, b]$. In order to "know" a number r such that $f(r)$ is the maximum value of f on $[a, b]$, we need only know a very small interval $[a', b']$ containing r. For example, if $b' - a' < 0.00005$, then the first four decimal digits of r' agree with those of a' (or b'). The first step in finding an interval smaller than $[a, b]$ that still contains r is to compute the midpoint $c = (a + b)/2$. Next we make a decision as to whether r lies in $[a, c]$ or in $[c, b]$. If $a \leq r \leq c$, then we replace b by c; otherwise, we replace a by c. Then the process repeats.

If you have studied Chapter 15, the preceding paragraph should look slightly familiar. However, there is an important difference between the preceding discussion and the use of interval-halving to find roots: We did not specify *how* the decision would be made to select either the left-hand or right-hand half-interval. Indeed, as we will see, a computer *cannot* make that decision. Nevertheless, the interval-halving algorithm just described leads to a sequence of intervals, each containing r, and contained in the previous one, and the length of the nth interval in the sequence is $(b - a)/2^n$. We will see that this provides a proof of the existence of a maximum value for f analogous to the proof of the Intermediate Value Theorem in Chapter 17. This development is completed in Chapter 23.

For practical purposes, we need to find a computable way to decide on an interval half the length of $[a, b]$ that contains the maximum value of f. It turns out that this can be done if we (1) restrict the class of functions under consideration and (2) increase the number of possible choices of half-intervals.

Figure 21.1

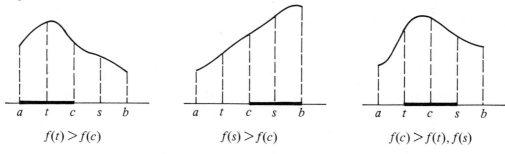

$$f(t) > f(c) \qquad\qquad f(s) > f(c) \qquad\qquad f(c) > f(t), f(s)$$

We restrict our attention now to a continuous function f on an interval $[a, b]$ such that f has no *relative minimum value* in the open interval (a, b). A moment's reflection will reveal that this means not only that the minimum value of f occurs at an endpoint, but also that there is exactly one relative maximum, namely, *the* maximum. (Why?) On the other hand, this restriction is not really serious if it is viewed in terms of "narrowing down" the given domain of f to an interval "of interest." In practice, this may mean some preliminary work to find an interval small enough to exclude interior relative minima. Note that the function of Problem 1 satisfies this restriction on its entire domain (as determined in Exercise 4).

The decision on selection of a half-interval is based on the values of f at three points: c, the midpoint t of $[a, c]$, and the midpoint s of $[c, b]$. Figure 21.1 illustrates three possibilities:

$$f(t) > f(c) \geq f(s),$$
$$f(s) > f(c) > f(t),$$

and

$$f(c) \geq \text{both } f(t) \text{ and } f(s).$$

Actually these are the *only* possibilities; $f(c) < \text{both } f(t)$ and $f(s)$ is not allowed, because this would require a relative minimum in the interval (t, s), and hence also in (a, b) (see Chap. 23 for additional details). In the first two cases, the choice of a half-interval is easy: If $f(t)$ is the largest of the three values, choose $[a, c]$; and if $f(s)$ is the largest of the three, choose $[c, b]$. The third case requires a new wrinkle: Because the maximum could be either left or right of c, neither $[a, c]$ nor $[c, b]$ would necessarily be a good choice. Instead, we choose the interval $[t, s]$, which also has the length $(b - a)/2$. You may want to make some other sketches like those in Figure 21.1 to convince yourself that this three-way decision process *always* selects a half-interval containing the maximum. The proof of this fact is deferred to Chapter 23.

The main part of this algorithm is summarized in the partial flowchart of Figure 21.2. As before, f designates a continuous function on $[a, b]$ with no interior relative minimum, and c, t, and s are computed from the formulas

$$c = \frac{a + b}{2}, \qquad t = \frac{a + c}{2}, \qquad s = \frac{c + b}{2}.$$

Figure 21.2

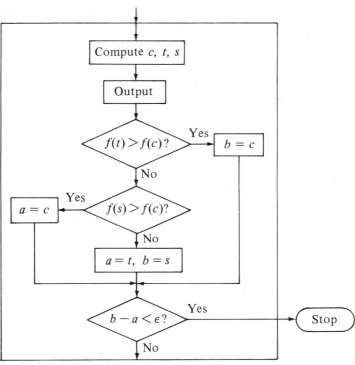

Exercise

6. Complete the flowchart, including definition of f, input of a and b, output of c and $f(c)$. Then write a program to implement the flowchart. Take $\epsilon = 0.00005$.

So far, we have ignored the problem of finding minima. Of course, we could repeat all the preceding steps with the inequalities reversed, and end up with a program that would minimize a function having no interior relative maximum. However, it is much simpler to just replace f by $-f$ and maximize (remembering that the final $f(c)$ will have the wrong sign).

Notice that our algorithm ensures four correct decimal places in the x-coordinate that locates the maximum value of f. However, this tells us nothing about how close the computed maximum value is to the true maximum value. This difficulty is the subject of another section of Chapter 23.

The next chapter presents several max/min problems that you can solve with the computer, using your program from Exercise 6. You should check whether the condition of no interior relative minimum is satisfied in each case (for f or $-f$, as appropriate), using your calculus techniques. Some of the problems involve trigonometric functions—if you have not yet studied their derivatives, look up the appropriate formulas in your calculus book. If you can't tell whether or not the condition is satisfied, experiment on the computer. But don't be satisfied with your answers unless you are sure they make sense.

[O]   A slicker version of the algorithm as previously described may be obtained by focusing on the midpoint of an interval of interest and on the half-length of the interval (rather than on endpoints, which were emphasized previously). Given a and b as before, compute initial values of $c = (a + b)/2$ (the midpoint), and $h = (b - a)/2$ (the half-length). The first step inside the loop is to replace h by $h/2$. If $f(c - h) > f(c)$, replace c by $c - h$. If $f(c + h) > f(c)$, replace c by $c + h$. If h is now small enough, exit from the loop. Otherwise, start over.

Exercise

7. [O] Check to see that the algorithm just described is equivalent to the previous one. (In particular, why is it unnecessary to take any specific action if the middle half-interval is selected?) Rewrite the program by using this version of the algorithm.

CHAPTER 22

COMPUTING EXTREME VALUES [L]

This session consists of several problems to be solved with your program from Exercise 6 of Chapter 21 (or Exercise 7, if you chose to do it that way). The first problem is included only for the purpose of checking out your program—you should be able to see the correct answer by inspection.

Exercises

1. Maximize $y = x - x^2$ on the interval [0, 3].

2. Solve the profit-maximizing problem of Chapter 21.

3. Maximize $f(x) = (x - 2)/(x^2 + 2x - 2)$ on the interval [2, 7].

4. Minimize the function of Exercise 3 on the interval $[-2.5, 0.5]$.

5. Minimize $x^5 + 4x^4 - x^3 + 3x^2 - x + 2$ on [0, 1].

6. Maximize sin $(3x^2)$ on [0, 1].

7. Maximize $\sin^2 x + \cos x$ on $[0, 2]$.

8. Minimize $\cos^2 (2x^2 - 1)$ on $[\sqrt{2}/2, \sqrt{2}]$.

9. Maximize $(x^2 - 6x + 9)^{1/3} - (9x^4 - 24x^3 + 16x^2)^{1/3}$ on $[1, 3]$.

NOTE: This function happens not to be differentiable throughout the indicated interval, although it is clearly continuous everywhere.

10. Maximize $x \sin x$ on $[2, 3]$.

11. Maximize $x^2 \sin x$ on $[2, 3]$.

12. Maximize $x^3 \sin x$ on $[2, 3]$.

13. Minimize $(\sin x)/x$ on $[4, 5]$.

The two remaining exercises require a root-finding program, such as the one you may have used with Chapters 16 or 19. If you have one handy, you might want to try these.

Exercises

14. The manufacturer with the problem of Exercise 2 might also be interested in knowing the root of the function f. This would be the break-even point beyond which spending more time per day on the process would actually result in a loss. Find the root.

15. While you are at it, you might check one of your max/min problems by finding a root of f'. For example, the derivative of the function in Exercise 6 is $f'(x) = 6x \cos (3x^2)$. Find a root between 0 and 1.

C H A P T E R   2 3

TYING UP SOME LOOSE ENDS [0]

In Chapter 21 we left undone three very important tasks: (1) the demonstration of the existence of maxima (and minima) based on interval-halving, (2) the proof that the three-way choice in our practical algorithm really does pick the right interval every time, and (3) the determination of how close the computed extreme value is to the true value. Each of these questions is taken up in this chapter.

23.1 EXISTENCE OF EXTREME VALUES

For the first task, we consider a continuous function f on a closed interval $[a, b]$, and we have to make a decision about whether the maximum value of f (if there is one) occurs in $[a, c]$ or in $[c, b]$, where $c = (a + b)/2$. In order to make the decision independent of whether or not f actually has a maximum value, we select $[a, c]$ if the following statement (*) is true, and select $[c, b]$ otherwise.

$\quad$ (*)There is an x in $[a, c]$ such that $f(x) \geq f(y)$ for every y in $[c, b]$.

Note that no computer can decide whether (*) is true or false, as this would require checking an infinite number of inequalities. Nevertheless, the statement is either true or false, and the interval-halving algorithm defines a sequence of intervals $[a_n, b_n]$, with $a_0 = a$, $b_0 = b$, and, in general,

$$a_{n+1} = a_n \quad \text{and} \quad b_{n+1} = \frac{a_n + b_n}{2} \quad \text{if (*) is true for } [a_n, b_n];$$

otherwise,

$$a_{n+1} = \frac{a_n + b_n}{2} \quad \text{and} \quad b_{n+1} = b_n.$$

Exercises

1. Show that the sequences $\{a_n\}$ and $\{b_n\}$ both converge. (HINT: Use Property 9.6.)

2. Show that the sequences $\{a_n\}$ and $\{b_n\}$ have the same limit L.

$\quad$ Our task now is to show that $f(L)$ is the maximum value of f on $[a, b]$; i.e., $f(x) \leq f(L)$ for every x in $[a, b]$.

$\quad$ Suppose that x_1 is in $[a, b]$, and $x_1 \neq L$. Then at some stage in the process of picking endpoints, an interval $[a_{n_1}, b_{n_1}]$ was chosen that did not contain x_1 but did contain L (because L is in every $[a_n, b_n]$). Because of (*), there is an x_2 in $[a_{n_1}, b_{n_1}]$ such that $f(x_2) \geq f(x_1)$. If $x_2 = L$, we have shown that $f(L) \geq f(x_1)$ where x_1 was arbitrary. If not, then there is some chosen interval $[a_{n_2}, b_{n_2}]$ that does not contain x_2; hence it contains a number x_3 such that $f(x_3) \geq f(x_2)$. Either $x_3 = L$, in which case $f(L) \geq f(x_2) \geq f(x_1)$, or $x_3 \neq L$, in which case we repeat the process to find x_4, and so on.

Exercises

3. If none of the x_k's is L, show that $\lim_{k \to \infty} x_k = L$, and that $\lim_{k \to \infty} f(x_k) = f(L)$. (HINT: Theorem 11.3.)

4. Show that $f(L) \geq f(x_1)$. Conclude that $f(L)$ is indeed the maximum value of f on $[a, b]$.

$\quad$ Thus our "noncomputable" interval-halving algorithm provides a proof of the important theorem that a *continuous function on a closed interval has a maximum* (and a minimum).

23.2 WHY DOES THE ALGORITHM WORK?

We turn now to the second task, that of showing that the computable three-way interval-halving algorithm actually makes the right choices. (The laboratory session of Chap. 22 should provide some empirical evidence for this.)

Let's review the bidding. We are considering a continuous function f on an interval $[a, b]$ such that f has no relative minimum in (a, b). We set $c = (a + b)/2$, $t = (a + c)/2$, and $s = (c + b)/2$. Our algorithm makes the following choices of half-intervals:

$$\begin{array}{ll} [a, c] & \text{if } f(t) > f(c) \geq f(s); \\ [c, b] & \text{if } f(t) \leq f(c) < f(s); \\ [t, s] & \text{if } f(c) \geq f(t) \text{ and } f(s). \end{array}$$

To show that the algorithm always works, we must prove

1. $f(c) < f(t)$ and $f(s)$ is impossible.

2. If $f(L)$ is the maximum value of f on $[a, b]$, then L lies in the chosen interval.

Exercises

5. Show: If f is continuous on $[x_1, x_2]$, and there is some z in (x_1, x_2) such that $f(z) < f(x_1)$ and $f(z) < f(x_2)$, then f has a (relative) minimum value on (x_1, x_2).

6. Show that, in the notation of the preceding algorithm, $f(c)$ cannot be less than both $f(t)$ and $f(s)$.

7. Suppose that $f(t) > f(c) \geq f(s)$, but L is not in the chosen interval $[a, c]$. Show that there must be a relative minimum on (a, b).

HINT: See Figure 23.1(a). Apply Exercise 5 with $x_1 = t$, $x_2 = L$.

8. Suppose that $f(t) \leq f(c) < f(s)$, but L is not in the chosen interval $[c, b]$. Show that there must be a relative minimum on (a, b) [see Fig. 23.1(b)].

Figure 23.1

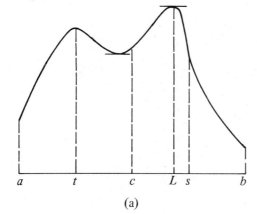
(a)

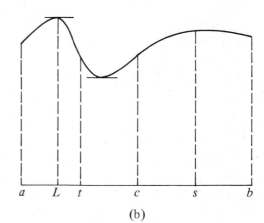
(b)

Figure 23.2

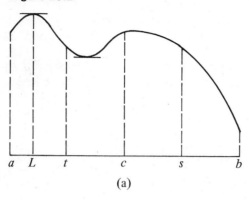

(a)

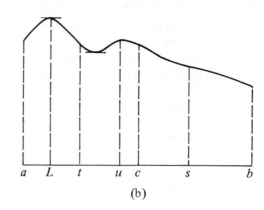

(b)

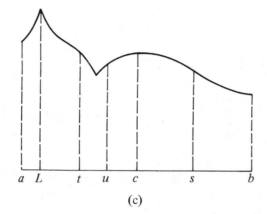

(c)

9. Suppose that $f(c) \geq f(t)$ and $f(c) \geq f(s)$, but L is not in the chosen interval $[t, s]$, say, $a \leq L < t$. (The other possibility is that $s < L \leq b$, and it may be treated similarly.) Show that there must be a relative minimum on (a, b).

HINT: See Figure 23.2. Consider separately the following cases:

  a. $f(t) < f(c)$.
  b. $f(t) = f(c)$ and $f(u) > f(t)$ for some u in (t, c).
  c. $f(t) = f(c)$ and $f(u) < f(t)$ for some u in (t, c).

Note that $f(u) = f(t)$ for $t < u < c$ implies that every point in (t, c) is a relative minimum point.

23.3 HOW WELL DOES IT WORK?

In our discussion in Chapter 21 of the interval-halving algorithm, we concentrated on determining a close approximation to a number r in $[a, b]$ so that $f(r)$ is the maximum value of f in $[a, b]$. If s were the computed approximation, we could control the size of $|s - r|$

by the interval-halving process, but no indication was given of the size of $|f(s) - f(r)|$. The algorithm required only that f be continuous on $[a, b]$, but we see in this section that assuming somewhat more about f leads naturally to an estimate of $|f(s) - f(r)|$. (The additional hypothesis is unnecessary, however, as you will recognize if you have already studied *uniform continuity*.) In particular, we observe how the Mean Value Theorem can be applied to a function f with continuous first derivative on an interval $[a, b]$ to conclude that everything there is to know about f—including its maxima and minima, as well as its roots—is contained in a *finite* list of values of f, indeed, a tabulation at equally spaced points. For practical purposes, we would not want to carry out many such tabulations, because the spacing between computed points might have to be very small. However, we will see that the *idea* of such a tabulation leads naturally to a (sometimes) practical method for deciding whether $|f(s) - f(r)| \leq \epsilon$ for some specified degree of accuracy ϵ. At the same time we will get an estimate for $|s - r|$, but this may be either smaller or larger than ϵ.

Let f be a function on $[a, b]$ that satisfies the preceding condition: $f'(x)$ is continuous on $[a, b]$. Because a continuous function on a closed interval is bounded, there exists a number $K > 0$ such that

$$|f'(x)| \leq K \qquad \text{for} \qquad a \leq x \leq b. \tag{23.1}$$

Exercises

10. For f and K as earlier, and any two numbers x_1 and x_2 in $[a, b]$, show that

$$|f(x_2) - f(x_1)| \leq K|x_2 - x_1|. \tag{23.2}$$

HINT: Use the Mean Value Theorem.

11. Given any positive number ϵ, set $\delta = \epsilon/K$, and imagine a tabulation of the values of f at the points $a, a + \delta, a + 2\delta$, and so on, up to b. Show that the maximum value $f(r)$ of f on $[a, b]$ must lie within ϵ of the largest tabulated value, and r must lie within δ of the number at which the tabulated value was computed.

Applying the idea of Exercise 11 to the algorithm of Chapter 21, we can think of our last computed interval, of length $\delta = (b - a)/2^n$, as part of a tabulation over the entire interval $[a, b]$ with spacing δ. If we know a bound K for the derivative, and we want to ensure accuracy of at least ϵ in the approximation to $f(r)$, we can change the end test to stop halving when $b - a < \epsilon/K$. (Here a and b mean the *last* computed endpoints, not the original ones.) The "Catch 22" aspect of this is that often the determination of the bound K is just as difficult a problem as the original maximization problem. Even when K can be found easily, the estimate of error that it provides may be very conservative, especially if the graph of f is very steep somewhere in the interval, but not necessarily near the extreme value. The latter problem can be reduced, however, by replacing the original interval by *any* smaller one that contains the extremum of interest, which should permit the choice of a smaller bound K.

Example   In Exercise 6 of Chapter 22, you were asked to maximize $f(x) = \sin(3x^2)$ on $[0, 1]$. Now $f'(x) = 6x \cos(3x^2)$ (see also Exercise 15 of Chap. 22); and because the cosine of anything is no more than 1 in absolute value,

$$|f'(x)| \le 6|x| \le 6 \qquad \text{for} \qquad 0 \le x \le 1.$$

Hence $K = 6$ bounds f', and your computation that determined the location of the maximum to within 0.00005 also gave the correct maximum value to within 0.0003. (The true answer for the maximum is 1, which is not surprising because the values of f are values of the sine function.)

Exercises

12. Consider the problem in Exercise 11 of Chapter 22, of maximizing $f(x) = x^2 \sin x$ on [2, 3]. Compute $f'(x)$, and verify that $K = 15$ will bound f' on the indicated interval. How many correct decimal places can you be sure of in your computed answer to this problem?

13. The discussion of this section can be used to construct an algorithm that will carry out a crude search for the maximum value of f on [a, b]. Given an error tolerance ϵ and a bound K for f', compute $\delta = \epsilon/K$, and have the computer "tabulate" f at $a, a + \delta, a + 2\delta$, and so on. To find the maximum, you need to keep track of only the largest computed value to date. When you reach the end of the tabulation, that will be the largest value, to within ϵ, of course. The "keeping track" is done as follows: Use a variable, say, M, for the largest value of f found so far; initially, M may be $f(a)$; each new value of $f(x)$ is compared with M, and whenever the new value is larger than M, M is replaced by that value. You will also want a separate variable for keeping track of the x-coordinate every time M changes. Make a flowchart for this search algorithm.

14. There would be little point in programming the algorithm of Exercise 13 if it were merely a very inefficient way to do what the interval-halving algorithm will do. But notice that it is not saddled with the "no interior minimum" restriction. Indeed, while keeping track of the largest value to date, we could also keep track of the smallest to date in the same way, thereby simultaneously computing the minimum of f on [a, b]. Furthermore, another pair of "keeping track" variables could be used to determine the smallest value of $|f|$ on [a, b]. If that turns out to be within ϵ of 0, you will have located a root of f, or if it does not, f has no root on [a, b]. Expand your flowchart to one for an algorithm that simultaneously searches for the maximum, minimum, and a root, and write a program to implement it. If you choose to try out this program, a value of $\epsilon = 0.01$ is appropriate. This search is necessarily crude because function evaluations numbering in the thousands (as might be the case here) take up noticeable amounts of computer time.

Remark   There is a technical name for inequality (23.2) that you may encounter in another context, perhaps in a later course in differential equations. A function f is said to satisfy a *Lipschitz condition* on [a, b] with *Lipschitz constant* K if

$$|f(x_1) - f(x_2)| \le K|x_1 - x_2|$$

for any two numbers x_1 and x_2 in $[a, b]$. Exercise 10 proves the following theorem: *If f' is continuous on $[a, b]$, then f satisfies a Lipschitz condition with K any bound for $f'(x)$ on $[a, b]$.*

References

Bell, et al.; Kuller; Schreiner.

C H A P T E R   2 4

UP THE UP STAIRCASE, OR FIRST STEPS IN INTEGRATION

24.0 HISTORICAL BACKGROUND [R]

In your calculus course you have, no doubt, learned to associate the abstract idea of (definite) integral of a function with the geometric idea of area bounded by a curve. The earliest nontrivial determination of curvilinear areas was carried out by a Greek mathematician of the fifth century B.C., Hippocrates of Chios,[1] who used properties of circles to compute areas of lunes (regions bounded by arcs of two different circles). During the fourth century B.C., Eudoxus of Cnidus worked out a Greek version of the integral calculus, which is now called the "method of exhaustion." He used his primitive limit ideas to prove theorems such as the proportionality of the area of a circle to the square of its diameter, or of circumference to diameter. The latter result was the basis of Archimedes' computation of π a century later.[2]

Archimedes also refined and developed the method of Eudoxus into a handy tool for solving all sorts of geometric problems, such as finding areas bounded by spirals, areas of parabolic segments, volumes of various solids, and surface area of the sphere. These are all routine exercises in calculus books now, primarily because of the insight of Newton and Leibniz that associated the evaluation of integrals (as limits of sums) with the reverse process of taking a derivative (indefinite integration), thereby obtaining a powerful computational tool called the Fundamental Theorem of Calculus.

---

[1]Not to be confused with his more celebrated contemporary, the physician Hippocrates of Cos.

[2]See Chapter 12.

But Archimedes had to compute his limits of sums the hard way, as did centuries of his mathematical successors. We have already noted in Chapter 12 the development of an integral calculus by the Japanese school of Seki Kowa at about the time of Newton and Leibniz. But their work more closely paralleled that of Archimedes than that of their European contemporaries.

There were some important steps preceding and hinting at the Fundamental Theorem. Your calculus book probably shows how to prove

$$\int_0^a x \, dx = \frac{a^2}{2}, \qquad \int_0^a x^2 \, dx = \frac{a^3}{3},$$

and perhaps even

$$\int_0^a x^3 \, dx = \frac{a^4}{4},$$

by using the limit-of-sums definition. At this point you are supposed to conclude (1) there *must* be an easier way to get these results, and (2) it's probably true for every positive integer n that

$$\int_0^a x^n \, dx = \frac{a^{n+1}}{n+1}. \tag{24.1}$$

(Thus you are pedagogically "set up" for the Fundamental Theorem.)

Formula (24.1) was first obtained by Bonaventura Cavalieri (1598–1647), who was a disciple of Galileo, a member of the Jesuate religious order, and a professor of mathematics at Bologna. He published the first few special cases in 1635 and the general result for positive integral powers in 1647. At about the same time, the French mathematician Pierre de Fermat (1601–1665) proved (24.1) by essentially the same method that appears today in calculus books (before introduction of the Fundamental Theorem). Later, however, Fermat gave another proof that extended the formula for integrating x^n to allow n to be fractional or negative ($\neq -1$). This involved a variation of the technique of upper and lower sums, using *unequal* intervals of subdivision whose lengths decrease in geometric progression. The problem of integrating x^{-1} was solved at about the same time by Gregory of St. Vincent (1584–1667), an itinerant Jesuit teacher. Born at Ghent (now in Belgium), he taught in Rome, Prague, and eventually at the court of Philip IV of Spain.

The world "integral" was coined in 1690 by Jacques Bernoulli,[3] who convinced Leibniz that the subject should be called *calculus integralis* rather than *calculus summatorius*. However, the seventeenth-century idea of integration was much more primitive and less sophisticated than that presented now in textbooks. Early in the nineteenth century, Augustin-Louis Cauchy (1789–1867) gave essentially the modern definition of the integral as a limit of sums, except that the function values were all taken at left-hand endpoints of the intervals of subdivision. His approach was guided by geometric intuition, and not entirely rigorous, but he knew that it was possible for discontinuous functions to have integrals.

---

[3]One of the more prominent members of a remarkable family that produced at least a dozen individuals who contributed to the development of mathematics over the period 1650–1850. See Boyer (1968, p. 456), for a family tree of the mathematical Bernoullis.

About 1850, what we now know as the *Riemann integral*, defined in terms of "Riemann sums," was introduced by Georg Friedrich Bernhard Riemann (1826–1866). He also established necessary and sufficient conditions for a bounded function to be integrable, and gave an example of a function that is integrable on a certain interval in spite of having infinitely many points of discontinuity in the interval.

24.1 UPPER AND LOWER SUMS

In standard calculus texts, once the Riemann integral (or a variation thereof) and the Fundamental Theorem have been introduced, the definition is seldom mentioned again (except possibly to show that certain physical or geometric situations really lead to integrals), and the student is encouraged to think of

$$\int_a^b f(x)\, dx = F(b) - F(a), \qquad \text{where } F'(x) = f(x), \tag{24.2}$$

as though *this* were the meaning of integral. An important part of the Fundamental Theorem is the fact that every continuous function f on $[a, b]$ *is* the derivative of something—i.e., there is another function F such that $F'(x) = f(x)$ for $a \le x \le b$. However, the theorem gives no clues about finding F in general, and the usefulness of (24.2) is obviously limited to situations in which an explicit expression for $F(x)$ is available. After you have spent perhaps several weeks studying techniques of "indefinite integration," you will discover that there are still many relatively simple continuous functions for which you do not know how to find an antiderivative. However, with a computer available, the large amount of arithmetic implicit in the limit-of-sums definition is not so intimidating, and the Fundamental Theorem may be bypassed when it is inconvenient or impossible to use. The purpose of the remainder of this chapter is to make a modest beginning at showing how to do just that.

Throughout the chapters on integration, we limit our attention to continuous functions. To establish a system of notation, we begin with a definition of the integral. This definition may or may not agree with the one in your calculus book, but it is equivalent, at least for continuous functions.

Definition Let f be a continuous function on $[a, b]$. For each positive integer n, let $x_0, x_1, \ldots, x_n$ be equally spaced points in the interval, so that $x_0 = a$, $x_n = b$, and, in general, $x_i = x_0 + i(\Delta x)$, where $\Delta x = (b - a)/n$. For each $i = 1, 2, \ldots, n$, let c_i be any number in the subinterval $[x_{i-1}, x_i]$. Form the sum

$$S_n = \sum_{i=1}^n f(c_i)\, \Delta x. \tag{24.3}$$

Then the *integral of f from a to b* is $\lim_{n \to \infty} S_n$, provided this limit exists for every choice of the c_i's and is independent of how the c_i's are chosen.

It is an important and nontrivial fact (i.e., a theorem) that the portion of the last sentence starting with "provided" is unnecessary: For continuous functions, the limit always exists and is independent of the choice of the c_i's, so that

$$\int_a^b f(x)\, dx = \lim_{n \to \infty} S_n, \tag{24.4}$$

where S_n is defined by (24.3).

For each fixed n, one way we can choose c_i in each $[x_{i-1}, x_i]$ is so that $f(c_i)$ is the minimum value of f on $[x_{i-1}, x_i]$. The resulting sum is called a *lower sum*, and is designated L_n (rather than S_n). Similarly, we could choose c_i to maximize f on each $[x_{i-1}, x_i]$. A sum corresponding to this choice is called an *upper sum*, and is denoted U_n. Because any choice of c_i produces a value of f between the minimum and maximum values, we have

$$L_n \leq S_n \leq U_n \qquad \text{for each } n, \tag{24.5}$$

where S_n is defined as in (24.3) for an arbitrary choice of the c_i's. Of course, all three sequences, L, S, and U, have the same limit, namely, $\int_a^b f(x) \, dx$.[4]

Another useful fact concerning upper and lower sums is

$$L_n \leq \int_a^b f(x) \, dx \leq U_n \qquad \text{for each } n. \tag{24.6}$$

In the case of a nonnegative function f, (24.6) is geometrically obvious because $\int_a^b f(x) \, dx$ is the area under the graph of f, and L_n and U_n are, respectively, areas of inscribed and circumscribed rectangular polygons. (We will not digress at this point to prove (24.6) in general.) The reason (24.6) is useful is that, if we compute upper and lower sums as approximations to the integral, then whenever $|U_n - L_n| < \epsilon$, both sums also differ from the integral by less than ϵ.

However, upper and lower sums also present a major difficulty: It is not obvious, in general, how to choose the c_i's. Indeed, maximizing or minimizing a continuous function on a closed interval is a substantial computational problem in its own right, as we have already seen; and it would be very inefficient to clutter up our integral computations with a large number of max/min problems (two for each subinterval for every n).

24.2 INTEGRALS OF MONOTONE FUNCTIONS

To avoid the problem just noted at the end of Section 24.1, for the rest of this chapter we restrict our attention to a special class of functions for which the max/min problem is always trivial: *monotone* functions, i.e., functions that are either increasing or decreasing throughout the interval $[a, b]$. (Recall that if f is differentiable on $[a, b]$, the derivative of the function provides an easy test for whether or not the function is monotone.) Clearly, if f is increasing from a to b, then the minimum value on $[x_{i-1}, x_i]$ occurs at x_{i-1}, and the maximum value occurs at x_i (and *vice versa* if f is decreasing). Hence the computation of upper and lower sums is equivalent (in either case) to the computation of *left-hand* and *right-hand* sums:

$$L_n = \sum_{i-1}^{n} f(x_{i-1}) \, \Delta x, \tag{24.7}$$

$$R_n = \sum_{i-1}^{n} f(x_i) \, \Delta x. \tag{24.8}$$

---

[4][O] If your calculus book defines the integral in terms of upper and lower sums, an application of the Pinching Theorem (9.7) to (24.5) shows that that definition is equivalent to the one given in this section.

(Note that we have changed the notation so that L now stands for "left" instead of "lower.")

Regardless of whether L_n or R_n is the larger of the two (i.e., of whether f is decreasing or increasing), the integral lies between them, so that $|L_n - R_n| < \epsilon$ implies that

$$\left| L_n - \int_a^b f(x)\, dx \right| < \epsilon \qquad \text{and} \qquad \left| R_n - \int_a^b f(x)\, dx \right| < \epsilon.$$

Everything in sight is now readily computable:

$$\Delta x = \frac{(b-a)}{n}, \qquad x_i = a + i(\Delta x), \qquad x_{i-1} = a + (i-1)\,\Delta x,$$

and we know how to accumulate the sums L_n and R_n (Sect. 4.2). In theory, all we have to do is compute L_n and R_n for $n = 1, 2, 3, \ldots$ and stop when $|L_n - R_n|$ is small enough. Unfortunately, that's not very practical, and for a rather obvious reason. A quick glance at one of those pictures in your calculus book of an approximation to an area by the area of a rectangular polygon (inscribed or circumscribed) shows that upper and lower sums don't really get close to the integral until n is quite large. Furthermore, our formulas for terms of the two sequences (24.7) and (24.8) are *explicit* rather than *recursive*, i.e., we can compute L_{1000}, say, without knowing any previous terms of the sequence. Hence if we know *which n* to use, we can approximate the integral by computing a single term of either sequence. This is not only a lot more efficient; it's also a lot easier to program.

Fortunately, it's rather easy in this situation to choose a suitable n. Consider what happens when we write out (24.7) and (24.8) and subtract

$$L_n = f(x_0)\,\Delta x + f(x_1)\,\Delta x + \cdots + f(x_{n-1})\,\Delta x$$

$$\underline{R_n = \qquad\qquad f(x_1)\,\Delta x + \cdots + f(x_{n-1})\,\Delta x + f(x_n)\,\Delta x}$$

$$L_n - R_n = f(x_0)\,\Delta x - f(x_n)\,\Delta x = [f(a) - f(b)]\,\Delta x.$$

All but one of the terms in L_n agree with all but one of the terms in R_n, so subtraction leaves only the difference of the first term of L_n and the last term of R_n. Hence

$$|L_n - R_n| = \frac{|f(a) - f(b)|\,(b-a)}{n}, \tag{24.9}$$

and this difference will be less than a specified tolerance ϵ if and only if

$$n > \frac{|f(a) - f(b)|\,(b-a)}{\epsilon}. \tag{24.10}$$

Given f, a, b, and ϵ, the right-hand side of (24.10) can be evaluated by the computer, and we can choose n to be any larger integer. For example, add 1 to the computed expression and use whatever version of the greatest-integer function is provided by your computer system.

Exercises

1. Make a flowchart for the evaluation method described in this section. A function f, assumed monotone, is to be defined in the program. Endpoints a and b are to be read in.

Take the end-test tolerance ϵ to be 0.005, as this method is rather crude. Compute an appropriate n to ensure that $|L_n - R_n| < \epsilon$. Then compute both L_n and R_n, as well as their average $T_n = (L_n + R_n)/2$. The output should include n, L_n, R_n, and T_n. Now write a program to implement the flowchart.

PROGRAMMING HINTS

a. In BASIC the function INT drops the fractional part of a number and returns the integer part.

b. The Δx in (24.7) and (24.8) is a constant factor in both sums. It is preferable to factor it out and multiply by it after accumulating a sum of function values. This is not only more efficient, but it also avoids adding up a large number of very small quantities, a procedure that may lead to unfortunate accumulation of roundoff errors. See Forsythe for a more thorough discussion of this problem.

c. Note that our previous determination of $L_n - R_n$ shows $R_n = L_n + [f(b) - f(a)] \Delta x$. Thus only L_n needs to be computed by accumulating a sum, and R_n can be computed from L_n.

2. [O] Observe that equation (24.9) implies that $\lim_{n \to \infty} |L_n - R_n| = 0$, and hence that L_n and R_n have the same limit. (Why?) Use this fact, together with (24.6), (24.5), and the Pinching Theorem (9.7), to prove that the integral of any monotone function on a closed interval exists and is the common limit of L_n and R_n. (Note that continuity is not even needed for this existence theorem.)

References

Boyer (1968); Forsythe; Gould; Grabiner; Kuller; Morgan.

C H A P T E R   2 5

INTEGRAL EVALUATIONS [L]

Each of the following exercises calls for the evaluation of a definite integral, using your program from Chapter 24. In each case, you should check that the indicated function is monotone on the appropriate interval. In those cases for which you have another method for finding the integral, check the number of correct decimal places in L_n, R_n, and T_n.

Exercises

1. $\int_0^1 x^4 \, dx.$ (Program checkout.)

2. $\int_{-1}^1 x^3 \, dx.$ (More program checkout.)

3. $\int_0^1 4\sqrt{1 - x^2} \, dx.$

NOTE: This integral represents the area of a circle of radius 1, computed as four times the area of the quarter-circle. That area is π, of course.

4. $\int_1^2 \sqrt{1 + x^4} \, dx.$

5. $\int_1^2 (1/x) \, dx.$

NOTE: This is a problem that you will learn to do via the Fundamental Theorem, if you have not already done so. But that method will not provide an explicit numerical answer as the method at hand does.

6. $\int_1^4 [x + (4/x)] \, dx.$

NOTE: The function $f(x) = x + (4/x)$ is *not* monotone. First express the integral as a sum of integrals over smaller intervals on which f is monotone.

7. $\int_0^b (\cos^3 x - \sin^3 x) \, dx$, where b is the smallest positive number such that $\sin^3 b = \cos^3 b$.

NOTE: You may be able to determine b by inspection. If not, use one of the root-finding programs from Chapter 15 or 18. This integral represents the area in the first quadrant bounded by $\sin^3 x$, $\cos^3 x$, and the y-axis. Later in your calculus course you may learn a method for integrating odd powers of sine and cosine.

C H A P T E R   2 6

TRAPEZOIDS: SECOND STEPS IN INTEGRATION

26.1 THE TRAPEZOIDAL RULE

Our treatment of definite integrals has, thus far, been restricted to consideration of monotone functions, for which we computed upper and lower sums (actually, left-hand sums L_n and right-hand sums R_n) and their averages (T_n). The reason for computing an average should be clear: If $|L_n - R_n| < \epsilon$, and the true value I of the integral lies between L_n and R_n, then $|I - T_n| < \epsilon/2$. (Why?) Hence we may reasonably expect that T_n will be a better approximation to I than is either L_n or R_n.

At least the first three exercises in Chapter 25 have alternative means of solution. (Depending on how far along you are in your calculus course, the last three may also have alternatives. In any case, your instructor knows how to compute accurate answers to those exercises by other methods.) If you have carefully compared your computed answers with known answers, you have already observed that the T_n's are indeed better approximations— not just by a factor of 2, as hinted earlier, but typically to twice as many correct decimal places (or more). We begin this chapter by examining (at an intuitive level) why that should be so. A more formal treatment is presented in Chapter 28.

One way to average the sums L_n and R_n is to average them term-by-term, and then sum the averages; i.e., from (24.7) and (24.8) we have

$$T_n = \sum_{i=1}^{n} \frac{f(x_{i-1}) + f(x_i)}{2} \Delta x. \tag{26.1}$$

This term-by-term averaging may be viewed geometrically, as in Figure 26.1, which illustrates the case of an increasing function f for $n = 4$. (The decreasing case is similar.) L_4 is the area of the inscribed rectangular polygon, R_4 the area of the circumscribed rectangular polygon, and T_4 is the area of another (circumscribed) polygon whose upper edges are *chords* of the graph of f, the ith chord joining the points $(x_{i-1}, f(x_{i-1}))$ and $(x_i, f(x_i))$. Because these chords fit the curve much better than do the stair steps, it should be no surprise that T_n approximates the integral much better than either L_n or R_n.

If we look at any one subinterval $[x_{i-1}, x_i]$ in Figure 26.1, we see that the portion of the area T_4 over that interval is the area of a *trapezoid* with "bases" $f(x_{i-1})$ and $f(x_i)$ and "height" Δx. (Rotate the page 90°, if necessary, to see this.) The ith term of the sum in formula (26.1) is precisely the area of such a trapezoid, and T_n is the sum of areas of

Figure 26.1

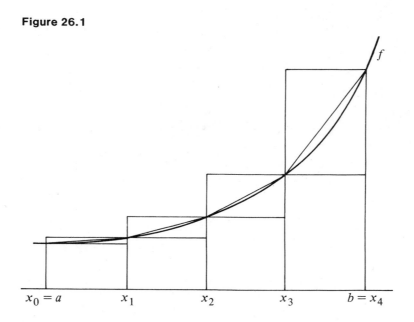

$x_0 = a$ x_1 x_2 x_3 $b = x_4$

trapezoids. This method of approximating integrals is usually called the *Trapezoidal Rule.* We arrived at formula (26.1) by consideration of a monotone function, and our intuitive discussion based on area implicitly assumed that the function is nonnegative as well. However, for any continuous function f on $[a, b]$, it is reasonable to approximate f on a sufficiently small subinterval $[x_{i-1}, x_i]$ by a linear function whose graph joins the points $(x_{i-1}, f(x_{i-1}))$ and $(x_i, f(x_i))$. The integral of the linear function from x_{i-1} to x_i is $\frac{1}{2}[f(x_{i-1}) + f(x_i)] \Delta x$, and the sum of these integrals is precisely T_n, as given by (26.1). Hence this formula should be a reasonable approximation to the integral in any case.

To make intelligent use of a formula like the Trapezoidal Rule, we need to know something about how the error $|T_n - \int_a^b f(x)\,dx|$ is related to n so that a suitable n can be chosen to achieve a desired accuracy in the answer. This is also related to the question of whether

$$\lim_{n \to \infty} T_n = \int_a^b f(x)\,dx, \tag{26.2}$$

which is evident for monotone functions. Actually, T_n always lies between the corresponding upper and lower sums, so (26.2) is correct for any continuous function. However, as with upper and lower sums, it is not very useful to compute many terms of the T_n sequence, and stop when two successive terms are close together. More precise information is needed.

26.2 THE MIDPOINT RULE

To obtain a reasonable error control for T_n, we turn our attention now to another, somewhat related, "trapezoidal" means of approximation. At this point, we need to assume that our functions are not merely continuous, but twice differentiable, so that tangent lines and concavity make sense throughout the interval $[a, b]$.

As usual, we subdivide $[a, b]$ into n equal subintervals, and on each we approximate the graph of f by a tangent line at the *midpoint* of the interval (see Fig. 26.2 for the case of $n =$

Figure 26.2

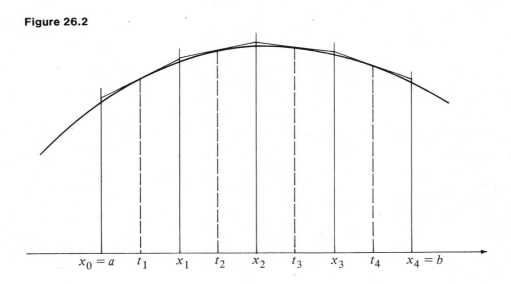

$x_0 = a \quad t_1 \quad x_1 \quad t_2 \quad x_2 \quad t_3 \quad x_3 \quad t_4 \quad x_4 = b$

Figure 26.3

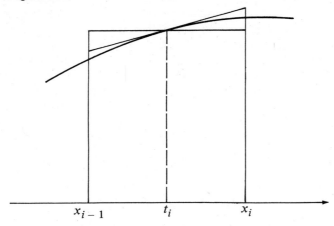

4). Our approximation to the integral is again obtained by summing the areas of trapezoids, but this time the "top" edge in each case is a segment of a tangent line, rather than a chord of a curve. The choice of the midpoint $t_i = (x_{i-1} + x_i)/2$ in each interval is convenient for computing the trapezoidal areas, because $f(t_i)$ is the "average base." Thus the area of the ith trapezoid is just $f(t_i) \, \Delta x$. (Equivalently, the area of the ith trapezoid is equal to the area of a rectangle of height $f(t_i)$ and width Δx—see Fig. 26.3.) This leads us to the following approximation to $\int_a^b f(x) \, dx$, called the *Midpoint Rule:*

$$M_n = \sum_{i=1}^{n} f(t_i) \, \Delta x, \tag{26.3}$$

where $t_i = (x_{i-1} + x_i)/2$.

It should come as no surprise that, as n gets larger, M_n more closely approximates the integral; i.e.,

$$\lim_{n \to \infty} M_n = \int_a^b f(x) \, dx. \tag{26.4}$$

A proof of this is sketched in Chapter 28. In the meantime if you pencil in a chord of the curve in Figure 26.3, you may come to the conclusion that the error involved in the midpoint approximation is likely to be smaller than that of the Trapezoidal Rule.

26.3 INTEGRALS OF CONCAVE FUNCTIONS

Let us now restrict our attention to functions f such that f'' is continuous and either positive throughout $[a, b]$ or negative throughout $[a, b]$. Thus the graph of f is either concave upward over the whole interval or concave downward, and there are no inflection points. (The two possibilities are illustrated by Figs. 26.1 and 26.2, respectively.) If $f''(x) > 0$ on $[a, b]$, then evidently chords of the graph lie *above* the graph and tangent lines lie *below* the graph. Thus

$$M_n \leq \int_a^b f(x) \, dx \leq T_n, \qquad \text{for each } n. \tag{26.5}$$

If $f''(x) < 0$, the situation is reversed: Tangent lines are always above the curve, and chords are below it, so

$$T_n \leq \int_a^b f(x)\,dx \leq M_n, \qquad \text{for each } n. \tag{26.6}$$

In either case, because the integral lies between M_n and T_n, whenever we have $|M_n - T_n| < \epsilon$, we also have

$$\left| M_n - \int_a^b f(x)\,dx \right| < \epsilon \qquad \text{and} \qquad \left| T_n - \int_a^b f(x)\,dx \right| < \epsilon.$$

Thus we have an effective means for controlling the error—i.e., for selecting an appropriate n to approximate the integral within ϵ, much as we did in Chapter 24—but this time much smaller values of n should work.

NOTE: Nothing in this paragraph depends on f being nonnegative, even though our illustrations were limited to that case.

For purposes of writing a program, we need to observe that there is a price to be paid for using better approximation methods: We *don't* have a simple formula for $|M_n - T_n|$ that would lead to a method for choosing n analogous to (24.10). Thus we have to compute terms of the M and T sequences until we find a pair of terms that differ by less than a specified tolerance. On the other hand, there really is little point in computing terms for $n = 1, 2, 3, \ldots$, because our basic formulas (26.1) and (26.3) are explicit, and we might as well get to large values of n more quickly. A reasonable compromise is to *double n* at each step instead of just adding 1. (Recall the computation of π in Chap. 13 that proceeded in this way.) Thus we can index a main loop by k, say, with $k = 1, 2, 3, \ldots$, and let $n = 2^k = 2, 4, 8, \ldots$. For each value of k, we can then compute the corresponding M_n and T_n, and test to see if their difference is small enough.

Within the main loop, we need sum-accumulating loops to evaluate T_n and M_n. Note that (26.1) can be rewritten as follows:

$$\begin{aligned}
T_n &= \frac{\Delta x}{2} \sum_{i=1}^{u} [f(x_{i-1}) + f(x_i)] \\
&= \frac{\Delta x}{2} \{[f(x_0) + f(x_1)] + [f(x_1) + f(x_2)] + \cdots + [f(x_{n-1}) + f(x_n)]\} \\
&= \frac{\Delta x}{2} [f(x_0) + 2f(x_1) + 2f(x_2) + \cdots + 2f(x_{n-1}) + f(x_n)]. \tag{26.7}
\end{aligned}$$

Programming T_n via (26.7) rather than (26.1) has the advantage that only about half as many function-evaluations are needed (these tend to use up a lot of computer time); but there is always a price to be paid for efficiency: The programming becomes slightly more complicated, because the first and last terms of the sum are different in form from the others. This complication can be minimized by rewriting the formula as

$$T_n = \Delta x \left[\frac{f(a) + f(b)}{2} + \sum_{i=1}^{n-1} f(x_i) \right]. \tag{26.8}$$

The preceding discussion is summarized in Figure 26.4, which shows a partial flowchart for the algorithm. The computations of T and M are to be done via formulas (26.8) and

Figure 26.4

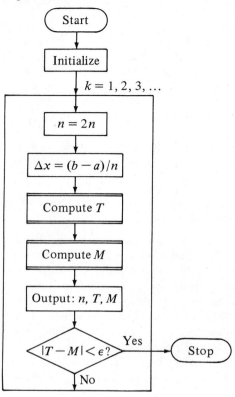

(26.3). You will also need

$$x_i = a + i(\Delta x), \qquad i = 0, 1, 2, \ldots, n, \qquad (26.9)$$

and

$$t_i = x_i - \frac{(\Delta x)}{2}, \qquad i = 1, 2, \ldots, n. \qquad (26.10)$$

Exercises

1. Complete the flowchart of Figure 26.4, including definition of f; input of a and b; initializing n; and the details of (26.8) and (26.3). Take $\epsilon = 0.00005$. Then write a program to implement the flowchart.

2. [O] The function values computed in evaluating T_n via (26.8) are all used again in the computation of T_{2n}. Rewrite (26.8) to get a recursive definition for the new T (i.e., T_{2n}) in terms of the old T (i.e., T_n) and some new function values. Incorporating this recursive definition of T into your program improves its efficiency by reducing the number of function evaluations. (HINT: The new Δx is *half* the old Δx.)

References

Hammer; Hart; Kuller.

C H A P T E R   2 7

INTEGRALS OF CONCAVE FUNCTIONS [L]

Each of the following exercises calls for evaluation of an integral by using your program from Chapter 26. In each case, check to see that the condition of the constant sign for $f''(x)$ is satisfied. Your end test won't work otherwise.

Exercises

1. $\int_0^2 x^4\, dx.$ (Program checkout.)
2. $\int_1^4 (x + 4/x)\, dx.$ (Compare Exercise 25.6.)
3. $\int_{-1}^1 2\sqrt{1 - x^2}\, dx.$ (What do you expect for an answer? Compare Exercise 25.3.)
4. $\int_0^2 \sqrt{1 + x^4}\, dx.$
5. $\int_1^2 1/(1 + x^2)\, dx.$
6. $\int_0^{\pi/2} 1/(1 + \cos x)\, dx.$
7. The function $1/(1 + x^2)$ of Exercise 5 is one that you will learn how to antidifferentiate in your calculus course (if you have not done so already). This function does *not* have constant concavity on the interval $[0, 1]$. Find the inflection point in that interval, and evaluate $\int_0^1 1/(1 + x^2)\, dx$ by writing it as the sum of two integrals.

NOTE: The value of the integral happens to be $\pi/4$. This is another way to evaluate π by integration.

8. $\int_1^3 \sqrt{2 + x^3}\, dx.$

9. $\int_0^1 (x^2 + x)^{3/2}\, dx.$

10. $\int_0^1 1/(x^3 + 7)\, dx.$

C H A P T E R 2 8

HOW GOOD IS THE TRAPEZOIDAL RULE? [O]

In this chapter we sketch a detailed derivation of the standard error estimate for the trapezoidal rule, a result that is often quoted in calculus books, but whose proof is invariably described as "beyond the level of this course." This result is *not* beyond the level of a beginning calculus course. On the contrary, it is a straightforward (but somewhat tedious) and useful application of the Fundamental Theorem of Calculus and other properties of the integral. We also sketch the parallel development of an error estimate for the Midpoint Rule. Throughout this chapter we assume that the functions under consideration have continuous second derivatives, but we do not require the concavity assumption made in Chapter 26.

The type of "error estimate" to be derived is an equality of the form

$$\left| T_n - \int_a^b f(x)\, dx \right| \le \frac{C}{n^2}, \qquad (28.1)$$

where T_n is the trapezoidal approximation given by (26.1) or (26.8), and C is a number depending only on the function f and the interval $[a, b]$, not on n. It is evident that (28.1) implies the convergence of the T_n sequence to the integral, but more importantly, it gives a means of determining a suitable n so that T_n approximates $\int_a^b f(x)\, dx$ to within a specified tolerance ϵ: n may be any integer such that

$$n > \sqrt{\frac{C}{\epsilon}}. \qquad (28.2)$$

The first step in deriving an error estimate is to cast the problem in as simple a form as possible. From Figure 26.1 we can see that the errors in the trapezoidal approximations on the individual subintervals may all be of the same sign. Thus it is reasonable to find an

Figure 28.1

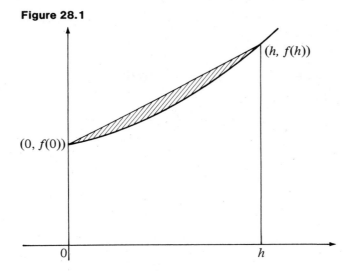

estimate for the error on a *single* subinterval that is independent of *which* subinterval is involved, and multiply by *n* to get the total estimate. [If the individual errors are *not* of the same sign, the total error may actually be much smaller than our estimate, but that will not affect the validity of a conclusion of the form (28.1).]

Figure 28.1 illustrates what appears to be a special case of our problem: determination of a bound on the difference between the trapezoidal area and the area under the graph of a function *f* on the interval [0, *h*]. We derive such a bound in a way that makes it depend on the length *h* of the interval, but not on position along the *x*-axis, so that the result may easily be translated to any interval of the form $[x_{i-1}, x_i]$ of length *h*. Let us begin by defining an *error function:*

$$E(h) = \frac{1}{2} h[f(0) + f(h)] - \int_0^h f(x)\, dx, \qquad h \geq 0. \tag{28.3}$$

Thus for each $h \geq 0$, $E(h)$ represents the shaded area in Figure 28.1. (But $E(h)$ might also be negative if the graph of *f* is not concave upward.) For future reference, note that

$$E(0) = 0. \tag{28.4}$$

Our problem is to find a bound for $|E(h)|$ in terms of *h*. The definition (28.3) happens to be too complicated to work with directly, but, as we will see, it leads to a rather simple expression for $E''(h)$. From a bound for the second derivative, we can then work back to the function $E(h)$, using the Fundamental Theorem and two important properties of integrals (proofs of which may be found in your calculus book):

$$\left| \int_a^b g(x)\, dx \right| \leq \int_a^b |g(x)|\, dx, \tag{28.5}$$

for any continuous function *g*; and:

$$\begin{aligned} &\text{if} \quad g_1(x) \leq g_2(x) \qquad \text{for } a \leq x \leq b, \\ &\text{then} \quad \int_a^b g_1(x)\, dx \leq \int_a^b g_2(x)\, dx. \end{aligned} \tag{28.6}$$

Exercises

1. Differentiate the error function E defined by (28.3) to obtain

$$E'(h) = \frac{1}{2}f(0) + \frac{1}{2}hf'(h) - \frac{1}{2}f(h).$$

Also note that

$$E'(0) = 0. \qquad\qquad (28.7)$$

HINTS: (a) $f(0)$ is a constant; (b) the Product Rule is needed for the term $\frac{1}{2} hf(h)$; (c) by the Fundamental Theorem of Calculus,

$$\frac{d}{dh}\int_0^h f(x)\,dx = f(h).$$

2. Differentiate again to obtain

$$E''(h) = \frac{1}{2}hf''(h).$$

Now recall that we assumed f'' to be a continuous function; hence f'' is bounded on any closed interval. Suppose K is a bound for f'' on *whatever* interval is of interest (we are being deliberately vague about the interval), i.e.,

$$|f''(t)| \le K \qquad \text{for all } t \text{ of interest.}$$

Then your computation of E'' shows that

$$|E''(t)| = \frac{1}{2}|tf''(t)| \le \frac{Kt}{2}, \qquad \text{for } 0 \le t \le h, \qquad (28.8)$$

provided K is a bound for $|f''|$ on $[0, h]$.

We are finally ready for the really interesting step: the "bootstrap" operation that allows us to convert the bound (28.8) for E'' first into a bound for E', then finally into a bound for E. Check the steps of the computation carefully:

$$|E'(h)| = |E'(h) - E'(0)|, \qquad \text{by (28.7),}$$

$$= \left|\int_0^h E''(t)\,dt\right|, \qquad \text{by the Fundamental Theorem,}$$

$$\le \int_0^h |E''(t)|\,dt, \qquad \text{by (28.5),}$$

$$\le \int_0^h \frac{Kt}{2}\,dt, \qquad \text{by (28.8) and (28.6),}$$

$$= \frac{Kh^2}{4}, \qquad \text{by the Fundamental Theorem.}$$

Replacing the variable h by t, we have

$$|E'(t)| \leq \frac{1}{4} Kt^2, \qquad \text{for } 0 \leq t \leq h. \tag{28.9}$$

Now we play the same game once more:

$$|E(h)| = |E(h) - E(0)|, \qquad \text{by (28.4)},$$

$$= \left| \int_0^h E'(t)\, dt \right|, \qquad \text{by the Fundamental Theorem},$$

$$\leq \int_0^h |E'(t)|\, dt, \qquad \text{by (28.5)},$$

$$\leq \int_0^h \frac{1}{4} Kt^2\, dt, \qquad \text{by (28.9) and (28.4)},$$

$$= \frac{Kh^3}{12}, \qquad \text{by the Fundamental Theorem}.$$

This is what we were looking for:

$$|E(h)| \leq \frac{Kh^3}{12}, \qquad \text{for } h \geq 0, \tag{28.10}$$

where K is any bound for $|f''(x)|$ on the interval of interest. (If you have successfully followed the argument to this point, perhaps you have gained a better understanding of why the Fundamental Theorem is called "fundamental.")

To complete our task of finding an inequality of the form (28.1), we now make a mental shift in point of view. Consider the function f on the closed interval $[a, b]$, and suppose K is a bound for $|f''(x)|$, $a \leq x \leq$ b. Let n be a fixed integer, and $x_0, x_1, x_2, \ldots, x_n$ be equally spaced points of subdivision in $[a, b]$. The preceding argument may be applied to each subinterval $[x_{i-1}, x_i]$ *as though* x_{i-1} were 0 and x_i were h, and hence $\Delta x = x_i - x_{i-1} = h$.[1] Then (28.10) tells us that the error in the trapezoidal approximation on the ith subinterval is no more than $K(\Delta x)^3/12$, and hence the total error is no more than n times this quantity. Replacing Δx by $(b - a)/n$, we have

$$\left| T_n - \int_a^b f(x)\, dx \right| \leq n \left(\frac{K}{12} \right) \left(\frac{b-a}{n} \right)^3 = \frac{K(b-a)^3}{12n^2}, \tag{28.11}$$

which has the desired form (28.1) with $C = K(b-a)^3/12$.

Example   Consider the problem of Exercise 2 in Chapter 27: Evaluate $\int_1^4 [x + (4/x)]\, dx$. How large should n be so that T_n is within 0.00005 of the true value of the integral? The function in question is $f(x) = x + (4/x)$, and you may check that $f''(x) = 8/x^3$. f'' is a nonnegative, decreasing function on $[1, 4]$, so its largest value (in absolute value) occurs at the left-hand endpoint. Thus $K = 8$ is a suitable bound. Then (28.11) tells us that a bound for

---

[1]More precisely, apply the argument to $f(x + x_{i-1})$ because $\int_{x_{i-1}}^{x_i} f(x)\, dx = \int_0^h f(x + x_i)\, dx$.

the error in T_n is

$$\frac{8 \cdot (4-1)^3}{12n^2} = \frac{18}{n^2}.$$

This number will be < 0.00005 if and only if

$$n > \sqrt{18/0.00005} = \sqrt{36 \cdot 10^4} = 600.$$

[See (28.2).] Thus, for example, $n = 601$ will do. Your experience at the terminal may show that this is rather conservative—in fact, $n = 256$ is sufficient to ensure four-decimal-place accuracy.

Exercises

3. Let $f(x) = x^4$.
   a. Show that $K = 48$ is a bound for $|f''(x)|$ on $[0, 2]$.
   b. What value of n will ensure that T_n is within 0.00005 of $\int_0^2 x^4\, dx$?
   c. How does this estimate fit with your computed results for Exercise 1 of Chapter 27?

4. Let $f(x) = 1/(1 + \cos x)$.
   a. Show that $f''(x) = (\cos x + \sin^2 x + 1)/(1 + \cos x)^3$.
   b. On the interval $[0, \pi/2]$, observe that $(1 + \cos x) \geq 1$, and hence $|f''(x)| \leq 3$.
   c. Show that $T_{150} - \int_0^{\pi/2} 1/(1 + \cos x)\, dx < 0.00005$. If you did Exercise 6 in Chapter 27, how does this compare with your computed result?

5. Write a program to evaluate $\int_a^b f(x)\, dx$ by a *single* application of the Trapezoidal Rule. The input is to include the endpoints a and b, the tolerance ϵ, and a bound K on $|f''(x)|$ for $a \leq x \leq b$. (K is to be computed by hand or, perhaps, by using an earlier program to maximize and minimize f'' on $[a, b]$.) The program should utilize (28.11) and (28.2) to compute an integer n such that T_n will lie within ϵ of the integral. Output should include n and T_n. If you have the opportunity to run this program, try it out with the exercises from Chapter 27.

In the remaining exercises, we sketch the derivation of an error estimate for M_n that is analogous to (28.11). As before, our first step is to look at the error on a single subinterval, chosen to make things as simple as possible. Because of the way M_n is evaluated, it is convenient to consider the interval $[-h/2, h/2]$ (see Fig. 28.2). We define the error function $E(h)$ as the difference between the trapezoidal area and the area under the graph of f:

$$E(h) = f(0)h - \int_{-h/2}^{h/2} f(x)\, dx$$
$$= f(0)h - \int_0^{h/2} f(x)\, dx + \int_0^{-h/2} f(x)\, dx. \tag{28.12}$$

Note that

$$E(0) = 0. \tag{28.13}$$

Figure 28.2

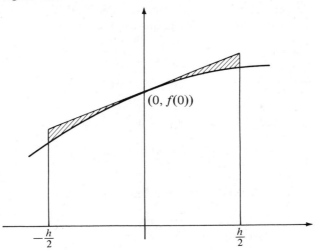

Exercises

6. Show that

$$E'(h) = f(0) - \frac{1}{2}f\left(\frac{h}{2}\right) - \frac{1}{2}f\left(-\frac{h}{2}\right) \tag{28.14}$$

and

$$E'(0) = 0. \tag{28.15}$$

NOTE: Don't forget to use the *Chain Rule* when differentiating the terms involving integrals.

7. Show that

$$E''(h) = \frac{1}{4}\left[f'\left(\frac{h}{2}\right) - f'\left(-\frac{h}{2}\right)\right]. \tag{28.16}$$

Then apply the Mean Value Theorem to f' on $[-h/2, h/2]$ to conclude that

$$E''(h) = \frac{h}{4}f''(c) \qquad \text{for some } c \text{ in } \left(-\frac{h}{2}, \frac{h}{2}\right). \tag{28.17}$$

Now we replace h by t, and it follows from (28.17) that, if K is a bound for $|f''(x)|$ on the interval of interest, then

$$|E''(t)| \le \frac{Kt}{4}, \qquad \text{for } t \ge 0. \tag{28.18}$$

NOTE: There is an important but subtle point here that did not come up in the previous discussion. Even though the "interval of interest" for f and its derivatives is of the form $[-h/2, h/2]$ for some (arbitrary) positive number h, the functions E, E', and E'' are defined on the interval $(0, \infty)$. Thus only nonnegative values of t need be considered in (28.18).

Exercises

8. Use (28.5), (28.6), (28.18), and the Fundamental Theorem to show that

$$| E'(h) | \le \frac{1}{8} Kh^2, \qquad \text{for } h \ge 0. \tag{28.19}$$

9. Change the variable in (28.19), and use a similar argument to show that

$$| E(h) | \le \frac{1}{24} Kh^3, \qquad \text{for } h \ge 0. \tag{28.20}$$

10. Replace h by $\Delta x = (b - a)/n$ and multiply the subinterval error by n to conclude that

$$\left| M_n - \int_a^b f(x)\, dx \right| \le \frac{K(b - a)^3}{24n^2}, \tag{28.21}$$

where K is a bound for $| f''(x) |$ on $[a, b]$.

11. Compute a value of n that will ensure M_n is correct to four decimal places for each of the problems considered in Exercises 3 and 4, and the example preceding Exercise 3.

Figure 28.3

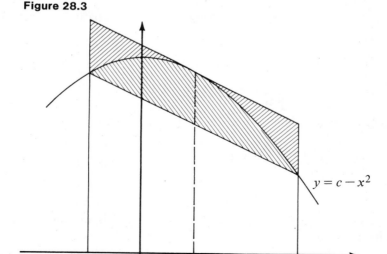

$$y = c - x^2$$

12. Write a program to evaluate $\int_a^b f(x)\,dx$ by a *single* application of the Midpoint Rule. (Compare Exercise 5.)

13. The error estimate (28.21) for M_n is exactly half the corresponding estimate (28.11) for T_n. Of course, this does not mean that the one error is half the other, in general. However, in the case of quadratic polynomial functions (parabolic graphs), the midpoint error is exactly half the trapezoidal error, as was known to Archimedes. By direct computation, prove the theorem of Archimedes for the case of $f(x) = c - x^2$ on the interval $[a, b]$:

$$\left| \frac{f(a) + f(b)}{2}(b - a) - \int_a^b f(x)\,dx \right| = 2 \left| f\left(\frac{a + b}{2}\right)(b - a) - \int_a^b f(x)\,dx \right|.$$

See Figure 28.3: The larger shaded area is twice the smaller.

Reference

Henriksen and Lees (Sect. 25).

C H A P T E R   2 9

PARABOLAS: THIRD STEPS IN INTEGRATION

29.0 A QUICK LOOK BACKWARD [R]

Let's review what we have done so far in computing (approximations to) definite integrals. Our first approach was to work directly from the definition and compute upper and lower sums. What this amounts to is approximating our given function for each small subinterval by a *constant* function (first one determined by the smallest value of f on the subinterval, then one determined by the largest value). The approximation to f obtained by piecing together all of these constant functions for a given subdivision of the interval $[a, b]$ is called a *step function* (because of the stair-step appearance of its graph), and the corresponding computed sum (lower or upper) is actually the integral from a to b of the approximating step function.

In Chapter 26 we moved on to better computational methods by using better approximations to f on each subinterval: *linear* approximations, rather than constant ones. (We considered the Trapezoidal and Midpoint Rules together in order to have a workable error control for functions of constant concavity, but if you have read Chapter 28, you know that each of these methods can be used independently of the other for any function with a continuous second derivative.) The approximating function obtained by piecing together all of the linear pieces for a given subdivision of $[a, b]$ is called *piecewise linear,* and the approximating trapezoidal sum is actually the integral from a to b of such a piecewise linear function.

Much of the practical application of calculus proceeds in just this manner: Complicated functions are approximated on suitably small portions of their domains by simpler functions, and the desired operations are performed on the simpler functions rather than on the more complicated ones. The simplest functions to work with are the polynomial functions, because (1) their values can always be obtained by simple arithmetic and (2) there are simple formulas for integrating and differentiating them.

Now that we have obtained integration methods by using approximating polynomial functions of degrees zero (constants) and one (linear functions), the next step in seeking greater accuracy (or better fit) is obviously to consider second-degree (quadratic) polynomials. In this chapter we do just that: We approximate f on suitably small subintervals by quadratics, piece these together to get a "piecewise quadratic" approximation over the interval $[a, b]$, and then approximate $\int_a^b f(x)\, dx$ by the integral of the approximating function, which is expressed in terms of a sum of values of f (and therefore is easily computable).

This procedure was first carried out by the Scottish mathematician James Gregory (a contemporary of Newton and Leibniz) and published in 1668, but in somewhat different form from that in use today. It was subsequently rediscovered by the self-taught British mathematician Thomas Simpson, who published in 1743 what we know today as Simpson's Rule, the formula that we are about to derive.

29.1 A QUADRATIC APPROXIMATION

The general form for a quadratic function is

$$p(x) = Ax^2 + Bx + C. \tag{29.1}$$

Thus to specify a quadratic, three coefficients A, B, and C must be determined, requiring (in general) three items of information (e.g., three points on its graph). Given a function f, we first consider the problem of finding a quadratic function p that agrees with f at three (equally spaced) points. For convenience, we assume that the x-values at which they must agree are $-h$, 0, and h for some positive number h. (We see later that there is really no loss in generality in considering this special case.) Our problem is illustrated by Figure 29.1, where we have set $y_0 = f(-h)$, $y_1 = f(0)$, and $y_2 = f(h)$. We are to find the function p of the form (29.1) such that $p(-h) = y_0, p(0) = y_1$, and $p(h) = y_2$. By direct substitution in (29.1), these three conditions become

$$Ah^2 - Bh + C = y_0,$$

$$C = y_1, \tag{29.2}$$

$$Ah^2 + Bh + C = y_2,$$

i.e., a set of three linear equations in the unknowns A, B, and C.

Figure 29.1

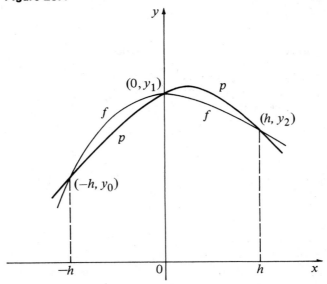

These equations are not difficult to solve by the method you learned in high school (see Exercise 4 that follows), but let's pause first to reflect on what it is we really want to know about p. Our objective is to express $\int_{-h}^{h} p(x)\, dx$ in terms of the given function values y_0, y_1, y_2 and the spacing h. (The resulting expression will then be our approximation to $\int_{-h}^{h} f(x)\, dx$.) From (29.1) we can compute directly

$$
\begin{aligned}
\int_{-h}^{h} p(x)\, dx &= \int_{-h}^{h} (Ax^2 + Bx + C)\, dx \\
&= \frac{Ax^3}{3} + \frac{Bx^2}{2} + Cx \Big|_{-h}^{h} \\
&= \frac{2Ah^3}{3} + 2Ch \\
&= \frac{h}{3}(2Ah^2 + 6C).
\end{aligned}
$$

Now we can save a few steps by observing from (29.2) that

$$y_0 + y_2 = 2Ah^2 + 2C$$

and

$$y_1 = C,$$

so

$$2Ah^2 + 6C = y_0 + y_2 + 4y_1.$$

Comparing this with our computation of the integral, we see that

$$\int_{-h}^{h} p(x)\, dx = \frac{h}{3}(y_0 + 4y_1 + y_2). \tag{29.3}$$

If y_0, y_1, and y_2 are positive, then (29.3) expresses the area under a parabolic arc in terms of three vertical distances and the horizontal spacing h between them. As such, this expression is independent of placement along the x-axis, so if our interval were centered at $x = c$ (instead of $x = 0$), we would have

$$\int_{c-h}^{c+h} p(x)\, dx = \frac{h}{3}(y_0 + 4y_1 + y_2), \tag{29.4}$$

where now $y_0 = f(c - h)$, $y_1 = f(c)$, and $y_2 = f(c + h)$. In fact, formula (29.4) is correct even if the y's are not all positive, but we will not bother to prove this.

29.2 SIMPSON'S RULE

Now let's recall that what we set out to do was to approximate $\int_a^b f(x)\, dx$, where f is any continuous function on $[a, b]$. As usual, we subdivide the interval into n pieces of equal length by points $x_0 = a$, x_1, x_2, ..., $x_n = b$, where $x_i - x_{i-1} = \Delta x = (b - a)/n$. We also require that the number n of subintervals be *even*. Then on each "double subinterval"

$$[x_0, x_2],\ [x_2, x_4],\ \ldots,\ [x_{n-2}, x_n],$$

we can approximate f by a quadratic that agrees with f at both endpoints and the midpoint (see Fig. 29.2). If we set $y_i = f(x_i)$ for $i = 0, 1, 2, \ldots, n$, then an approximation to the integral of f on $[x_{i-1}, x_{i+1}]$ is given, according to (29.4), by

$$\frac{\Delta x}{3}(y_{i-1} + 4y_i + y_{i+1}).$$

Figure 29.2

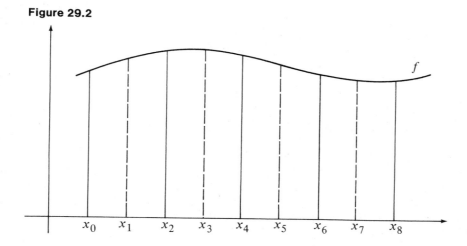

These are now added up for $i = 1, 3, 5, \ldots, n - 1$, to produce the following approximation for $\int_a^b f(x)\, dx$:

$$\frac{\Delta x}{3}(y_0 + 4y_1 + y_2) + \frac{\Delta x}{3}(y_2 + 4y_3 + y_4) + \cdots + \frac{\Delta x}{3}(y_{n-2} + 4y_{n-1} + y_n).$$

We may factor out $\Delta x/3$ and combine like terms (involving $y_2, y_4, \ldots, y_{n-2}$) to arrive at the final form of *Simpson's Rule:*

$$\int_a^b f(x)\, dx = \frac{\Delta x}{3}(y_0 + 4y_1 + 2y_2 + 4y_3 + 2y_4 + \cdots + 2y_{n-2} + 4y_{n-1} + y_n), \quad (29.5)$$

where $\Delta x = (b - a)/n$, and $y_i = f(a + i\Delta x)$ for each i.

Of course, in order to use (29.5) intelligently, we have to consider the usual problem of deciding how large n should be to achieve some desired accuracy with the approximation. Intuition suggests that quadratic-curve fits ought to be substantially better than linear ones. (Why else go to all this trouble?) But this time we don't have any clear-cut way of deciding when n is large enough. Let's abbreviate the right-hand side of (29.5) by S_n (the nth Simpson approximation to the integral), and write E_n for the error in this approximation:

$$E_n = \left| \int_a^b f(x)\, dx - S_n \right|. \quad (29.6)$$

By a process similar to that carried out for the Trapezoidal and Midpoint Rules in Chapter 28 (but about twice as long and complicated), one may show that

$$E_n \leqslant \frac{K(b - a)^5}{180n^4}, \quad (29.7)$$

where K is a bound for values of the fourth derivative of f on $[a, b]$. (The proof of (29.7) is outlined in Chap. 31.) Later we give some examples of how the estimate (29.7) can be used, but it is often the case that a suitable K is somewhat difficult to determine, and therefore use of the estimate for determining n is not really practical.

An alternative approach is to compute terms of the sequence $S_2, S_4, S_6, \ldots$ (remember that n has to be even!), and stop when we get "convergence" in some sense, since it is clear from (29.6) and (29.7) that

$$\lim_{n \to \infty} S_n = \int_a^b f(x)\, dx.$$

There are two problems with this, one of which we have observed before (Chap. 26): (1) If the "right" n is on the order of 1000, say, then we have to evaluate Simpson's Rule about 500 times to get there and (2) we don't have anything with which to compare S_n to decide when it is close enough to the true value of the integral.

For the Trapezoidal Rule, we overcame problem (1) by *doubling* n at each step, i.e., computing an approximation for $n = 2, 4, 8, 16, \ldots$, and we may as well do the same thing here. (Note that $2^{10} = 1024$, so it takes only ten steps to get to numbers that large, which should be adequate for all practical purposes.) For problem (2), we just have to admit defeat as far as a mathematically complete, correct, and available solution is concerned, and instead fall back on a rule-of-thumb of experienced computer users: When there is reason to believe that a sequence converges rapidly, if two successive terms are close to each other, they are also likely to be close to the limit. (If you have studied Chap. 18, you have seen a convergence test like this before but that was a situation in which we were able to prove that it gave correct answers.) Thus we will stop the computation when

$$| S_{2^k} - S_{2^{k+1}} |$$

is small enough (less than ϵ), and *hope* that each of these numbers lies within ϵ of the right answer. Our reason for believing that the sequence converges rapidly (at least for functions with continuous fourth derivatives) is (29.7), which indicates that the error should diminish rapidly as n increases.

WARNING: This would not necessarily work if we were to compute all the terms S_2, S_4, S_6, S_8, ... If an "optimal" n were on the order of 500, say, one would expect S_{208} and S_{210} to be very close to each other, even if neither were close enough to the right answer. This is another reason for doubling n at each step. Too much computation may lead to wrong answers!

Exercise

1. The partial flowchart in Figure 29.3 shows how S may be computed from formula (29.5) and then compared with the last computed value of S, saved as T before returning to the

Figure 29.3

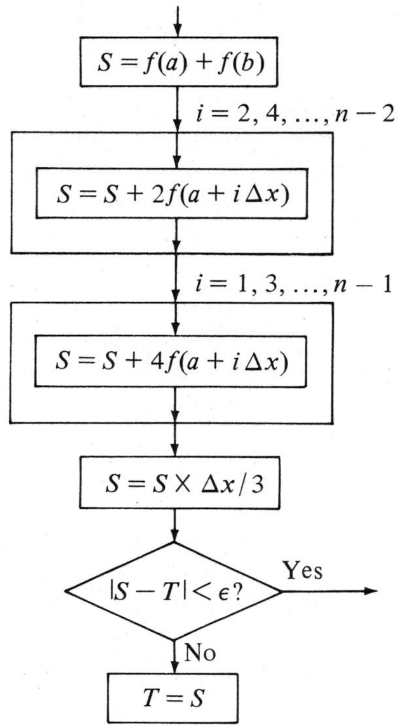

top of the main loop (not shown). After checking to make sure you understand this part of the algorithm, complete the flowchart to include the necessary initialization, doubling n, output, and so on. (Note that T has to be initialized before entering the main loop, in such a way that the end test will never be satisfied the first time through.) Leave ϵ variable, to be input at execution time.

29.3 USING THE ERROR ESTIMATE [O]

In this section we consider some examples of what can and cannot be done with the error estimate (29.7).

Example Suppose we want to find $\int_0^{\pi/2} \sin x \, dx$ to within 0.00005. (Note that this problem can be solved more easily by hand calculation than by Simpson's Rule—its only value as an illustration is that we can use it to check out the program when we are ready to run it.) If $f(x) = \sin x$, then the fourth derivative is also $\sin x$ (why?), and

$$|f^{(4)}(x)| \leq 1$$

for all x. The error estimate (29.7) tells us that

$$E_n \leq \frac{(\pi/2)^5}{180n^4}.$$

Thus we will have $E_n < 0.00005$ if

$$n^4 > \frac{(\pi/2)^5}{180(0.00005)}.$$

One may check that $n = 8$ is large enough, by taking a crude upper bound for $\pi/2$ (say, 1.6) and doing a little arithmetic. (Do it!)

Exercises

2. Suppose we want to evaluate $\int_1^4 [x + (4/x)] \, dx$ to within 0.00001. (This integral appeared in Exercise 6 of Chap. 25 and Exercise 2 of Chap. 27.)
   a. Verify that $f^{(4)}(x) = 96x^{-5}$ and that $K = 96$ will do for a bound for $f^{(4)}$ on $[1, 4]$.
   b. Use the error estimate (29.7) to verify that the desired accuracy is achieved whenever $n \geq 60$.

3. Consider the integral of Exercise 5 of Chapter 27:

$$\int_1^2 \frac{1}{1 + x^2} \, dx.$$

   a. Verify that

$$f^{(4)}(x) = \frac{24(1 - 10x^2 + 5x^4)}{(1 + x^2)^5}.$$

(This is a good test of your skill at computing derivatives of rational functions.)

b. Observe that $(1 + x^2)^5 \geq 1$ on $[1, 2]$.

c. Use your calculus techniques to maximize $1 - 10x^2 + 5x^4$ on $[1, 2]$.

d. Show that $K = 984$ will bound $f^{(4)}$ on $[1, 2]$. (This is possibly a very conservative bound because we have deliberately avoided the much more difficult task of finding the extreme values of $f^{(4)}$ on $[1, 2]$.)

e. Show that the error in S_{16} is less than 0.0001.

These examples, which have been carefully selected to avoid really difficult calculations, illustrate that it may be possible to select a single n to calculate an integral by Simpson's Rule to within a desired accuracy. However, it will often be the case that finding a suitable bound K is a formidable task, and it is much simpler (and it usually works) to compute a *sequence* of approximations that stops when consecutive terms are sufficiently close, as previously indicated.

Exercises

4. Solve the equations (29.2) for the coefficients A, B, and C of the quadratic function whose graph passes through $(-h, y_0)$, $(0, y_1)$, and (h, y_2). [ANSWER:

$$A = \frac{1}{2h^2} (y_0 - 2y_1 + y_2), \qquad B = \frac{1}{2h} (y_2 - y_0), \qquad C = y_1.]$$

5. All the function values used in computing S_{2^k} are used again in computing $S_{2^{k+1}}$. Is it feasible to avoid recomputing these function values by expressing $S_{2^{k+1}}$ recursively in terms of S_{2^k} and new function values?

References

Monzino; Munro; Peters and Maley.

C H A P T E R 3 0

INTEGRALS VIA SIMPSON'S RULE [L]

The following exercises are to be solved with your program from Chapter 29. End-test tolerance on the order of 0.00005 will work in all cases. For integrals on intervals of length 1 and for which the integrand clearly has a bounded fourth derivative, you might experiment with tolerances ranging down to 10^{-6}. Note that in some cases the integrand is *not* differentiable throughout the interval, let alone having a bounded fourth derivative. Nevertheless, Simpson's Rule often works without any theoretical support. Also note that some of the problems require prior determination of the function to be integrated (e.g., the arclength problems).

Exercises

1. Evaluate $\int_0^{\pi/2} \sin x \, dx$. (Program checkout.) According to the example in Section 29.3, $n = 8$ should be sufficient to achieve four-place accuracy. Is it?

2. Evaluate $\int_1^4 [x + (4/x)] \, dx$. (According to Exercise 29.2, S_{64} should be within 10^{-5} of the correct answer.)

3. Evaluate $\int_1^2 1/(1 + x^2) \, dx$. (According to Exercise 29.3, S_{16} should be within 10^{-4} of the correct answer.)

4. Compute π as the area of a circle of radius 1: $\int_0^1 4\sqrt{1 - x^2} \, dx$.

5. Evaluate $\int_0^2 (x^2 + x + 7)^{3/2} \, dx$.

6. Evaluate $\int_{-1}^1 1/(x^3 + 2) \, dx$.

7. Evaluate $\int_0^2 (x^2 - 1)/(x^2 + 1) \, dx$.

8. Compute the arclength of the graph of $y = 1/x$ between $(1, 1)$ and $(5, \frac{1}{5})$.

9. Compute the arclength of the graph of $y = x^{5/2}$ between 0 and 1.

10. Find the total area bounded by the graph of $y = x - \cos x$ and the x-axis between $x = 0$ and $x = 1$. (See Fig. 30.1.) Note that the calculus method for doing this requires finding where $y = 0$ and evaluating *two* integrals. You can skip the root-finding step and evaluate $\int_0^1 |x - \cos x| \, dx$ (see Fig. 30.2).

11. Exercise 10 is one in which the integrand is clearly not differentiable; therefore we have no reason to believe that Simpson's Rule actually works. Check the calculation by doing it the other way, as the sum of two integrals. (The necessary root is the answer to Exercise 10 in Chap. 19.)

Figure 30.1

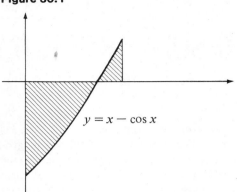

$y = x - \cos x$

Figure 30.2

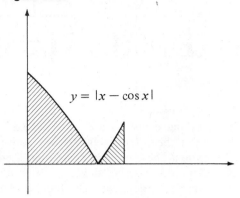

$y = |x - \cos x|$

12. Calculate π as an arclength.

NOTE: π is the length of the semicircle $y = \sqrt{1 - x^2}$ from -1 to 1. However, dy/dx is not continuous at the endpoints, so the arclength formula does not work for this arc. One possibility is to compute twice the length of the quarter-circle $y = \sqrt{1 - x^2}$ from $-\sqrt{2}/2$ to $\sqrt{2}/2$.

Reference

Coolidge (1953).

CHAPTER 31

HOW GOOD IS SIMPSON'S RULE? [O]

Our purpose in this chapter is to demonstrate how one obtains an error estimate of the form (29.7) for Simpson's Rule. This expression for the error was first published by the British mathematician James Stirling (1692–1770) in 1730, more than a decade before Simpson published his "rule"! The procedure is similar to that followed in Chapter 28: We introduce an error function, we differentiate it several times, we find a bound for one of its derivatives, and then we work back up to a bound on the error function itself by successive applications of the Fundamental Theorem. This process is carried out on one "double subinterval," and then the estimate is multiplied by $n/2$ to get a bound on the total error.

It should be noted that this kind of error estimate requires knowledge of the function being integrated in an explicit analytic form in order to be useful. However, Simpson's Rule (and other such formulas) can be used to integrate functions given in tabular form, by experimental data, or in any manner that provides enough equally spaced values. When nothing is known about the function except the given values, the problem of estimating error is quite different in nature. This kind of estimate was first carried out for Simpson's Rule by J. B. Scarborough in 1926 (see Bibliography).

Error estimates that use a bound on a certain derivative over the entire interval of integration and that assume that the error on the worst subinterval will be repeated throughout the interval are necessarily conservative. (You may have observed in your last two laboratory sessions that the desired accuracy, when you knew the right answer, could sometimes be achieved much sooner than the theory predicted.)

We begin as we did in Chapter 28: Our problem is to determine a bound for the difference between $\int_{-h}^{h} f(x)\,dx$ and the corresponding integral for the quadratic function whose graph passes through the points $(-h, f(-h))$, $(0, f(0))$, and $(h, f(h))$ (see Fig. 29.1). The latter integral, according to (29.3), is given by $(h/3)\,[f(-h) + 4f(0) + f(h)]$. Thus our error function is given by:

$$E(h) = \frac{h}{3}\,[f(-h) + 4f(0) + f(h)] - \int_{-h}^{h} f(x)\,dx, \qquad h \geqslant 0. \tag{31.1}$$

Exercise

1. Using the same differentiation rules as those used in Chapter 28, show that the first three derivatives of $E(h)$ are given by

$$E'(h) = \frac{h}{3}\,[f'(h) - f'(-h)] - \frac{2}{3}f(h) - \frac{2}{3}f(-h) + \frac{4}{3}f(0); \tag{31.2}$$

$$E''(h) = \frac{h}{3}\,[f''(h) + f''(-h)] - \frac{1}{3}f'(h) + \frac{1}{3}f'(-h); \tag{31.3}$$

$$E'''(h) = \frac{h}{3}\,[f'''(h) - f'''(-h)]. \tag{31.4}$$

Also check that

$$E(0) = E'(0) = E''(0) = 0. \tag{31.5}$$

The difference of function values in (31.4) can be replaced, according to the Mean Value Theorem, by a value of the derivative of f''' times the length of the interval—i.e.,

$$E'''(h) = \frac{h}{3}\,[2hf^{(4)}(c)] = \frac{2h^2}{3}f^{(4)}(c), \tag{31.6}$$

for some c in $(-h, h)$. Now suppose that $K \geq |f^{(4)}(x)|$ for all x in the interval of interest. If we replace h by t in (31.6), it follows that

$$| E'''(t) | \leq \frac{2}{3} t^2 K, \qquad \text{for } t \geq 0. \tag{31.7}$$

Now we are ready for the "bootstrap" operation, which proceeds exactly as in Chapter 28, except that we are now starting with E''' instead of E''. Check the steps carefully:

$$| E'''(h) | = | E''(h) - E''(0) | \qquad \text{by (31.5)}$$

$$= \left| \int_0^h E'''(t) \, dt \right| \qquad \text{by the Fundamental Theorem}$$

$$\leq \int_0^h | E'''(t) | \, dt \qquad \text{by (28.5)}$$

$$\leq \int_0^h \frac{2}{3} Kt^2 \, dt \qquad \text{by (31.7) and (28.6)}$$

$$= \frac{2}{9} Kh^3 \qquad \text{by the Fundamental Theorem.}$$

Replacing h by t again, we have

$$| E''(t) | \leq \frac{2}{9} Kt^3, \qquad \text{for } t \geq 0. \tag{31.8}$$

Exercise

2. Using the same steps as before, show that

$$| E'(t) | \leq \frac{1}{18} Kt^4, \qquad \text{for } t \geq 0. \tag{31.9}$$

Then repeat the process once more to show that

$$| E(h) | \leq \frac{1}{90} Kh^5, \qquad \text{for } h \geq 0. \tag{31.10}$$

Inequality (31.10) expresses a bound for the possible error in using a quadratic approximation on each of $n/2$ subintervals of $[a, b]$, where $h = \Delta x = (b - a)/n$. Thus the total error E_n defined by (29.6) is bounded by

$$\frac{n}{2} \frac{K(\Delta x)^5}{90} = \frac{nK(b - a)^5}{180n^5} = \frac{K(b - a)^5}{180n^4},$$

which completes the proof of (29.7).

One consequence of this error estimate is a little surprising. If f is a cubic polynomial function, then $f^{(4)}(x) = 0$ for all x. Hence we may take $K = 0$ as a bound for the fourth derivative. It follows that the error in approximating the integral of a third-degree

Figure 31.1

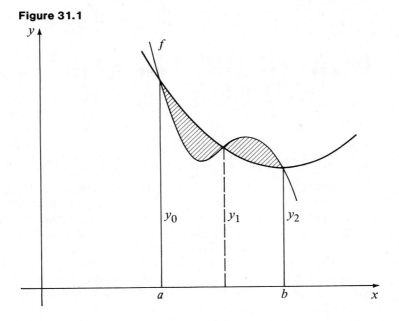

polynomial by Simpson's Rule (even with $n = 2$) must be 0. This situation is illustrated in Fig. 31.1: The two shaded areas must be equal!

References

Henriksen and Lees (Sect. 25); Scarborough.

CHAPTER 32

HOW TO DIFFERENTIATE, AND HOW NOT TO

32.0 HISTORICAL NOTE [R]

The subject of derivatives and differentiation is intimately tied up with the idea of a tangent line to a graph, or of best linear approximation to a function at a single point. The origins of this subject are not quite so ancient as those of integration (see Chap. 24), but the story again begins in Greece, this time with a younger rival of Archimedes, Appolonius of Perga (ca. 263–ca. 190 B.C.). In Book V of his work *Conics,* Appolonius determined tangent and normal lines to conic sections (parabolas, ellipses, and hyperbolas) in terms of maximum or minimum properties of these lines (e.g., the normal from a point P to a curve C was defined as the line joining P to a point Q on the curve for which the distance PQ was either a maximum or a minimum).[1] In the seventeenth century, the Frenchman Pierre de Fermat (1601–1665) reconstructed the work of Appolonius and developed an analytic geometry similar to that in use today. (His contemporary, René Descartes, is usually given credit for discovering analytic geometry and is remembered in the term *Cartesian* coordinates. Fermat's work was done at least as early but was not published until much later.)

In the early 1630s, Fermat showed how to compute maxima and minima for polynomials by a method that amounted to setting a limit of a difference quotient equal to zero. He also computed slopes of tangent lines by a similar method but was not able to give a satisfactory explanation of his method. Of course, Fermat did not speak in terms of taking limits of difference quotients. His process was more like finding a factor of Δx to cancel with the denominator and then setting all remaining Δx's to zero (does that sound familiar?). The idea of canceling factors secretly known to be zero raised some eyebrows, even though it gave right answers, but it was not until the mid-eighteenth century that Jean LeRond d'Alembert (1717–1783) first expressed the process of differentiation explicitly in terms of finding limits of difference quotients. (Besides being a mathematician, d'Alembert was thoroughly educated in a wide variety of fields and was coauthor with Denis Diderot of the celebrated 28-volume *Encyclopédie.*) In the 1820s, another Frenchman, Augustin-Louis Cauchy (1789–1867), developed d'Alembert's idea of differentiation by giving a precise arithmetic definition of limit. This was published in an elementary calculus book that had a great deal to do with shaping the subject as we know it today. (Work similar to Cauchy's was done independently by Bernhard Bolzano (1781–1848), a Czech priest, but his work was largely unnoticed by his contemporaries.)

---

[1]See also Coolidge (1951), who asserts that the Appolonian theorems must have been known sometime before Appolonius, in particular, to Euclid, about 300 B.C.

32.1 WHAT'S WRONG WITH CALCULUS TECHNIQUES?

In an early laboratory session, we experimented with computation of derivatives as limits of difference quotients (Chap. 10). In the meantime you have learned how to combine the Product and Quotient Rules, the Power Rule, the Chain Rule, and perhaps other rules, to differentiate all sorts of complicated functions without referring to the original definition. For example, you would have no difficulty writing down the derivative of

$$f(x) = \frac{(x^{5/2} + 2x^2 + x)^{3/4} (\sqrt{x^3 + 1} + x^7)^{1/3}}{(14x^3 + 12x^2 + \sqrt{x} + 29)^{2/7}}, \tag{32.1}$$

right? On the other hand, it might be nice to have a little help with such a job.

You may have wondered whether it is possible to use a computer to manipulate expressions like (32.1) "symbolically" and produce the derivative function in similar form, according to the rules of elementary calculus. Each rule like this is a simple algorithm, and it is not too difficult to combine them into a program. The hard part is to teach the computer how to simplify the result, a task that is much easier to *do* than it is to describe *how* to do because you can "see" things such as "like terms" that would be very hard for the computer to "see." If we limited our attention to polynomials, it might not be too difficult to write a differentiation program, but it would solve only problems that are easy to solve with pencil and paper. In fact, many man-years of effort by computer scientists have been invested in the development of symbolic manipulation routines that incorporate, among other things, all the formulas, computational techniques, and simplifications that are taught in courses in algebra, trigonometry, and calculus.

An example of such a system is MACSYMA (for Project MAC SYmbolic MAnipulation program), which was developed at MIT and is now available on a small number of large computers. Educational institutions can access MACSYMA via interactive terminals with telephone connections to EDUNET, a nationwide network for sharing educational computer resources. However, MACSYMA and EDUNET are much too expensive to turn over to every calculus student for solving homework problems.

A more recent development is a symbolic manipulation system called muMATH, which has many (but not all) of the capabilities of MACSYMA and runs on (some) microcomputers. In particular, muMATH *can* differentiate the function defined by (32.1), as well as any others that are likely to turn up in a calculus course, and it can simplify the answers as well. We are not too far from the day when handheld devices that are not much bigger than today's calculators will have the capabilities of muMATH, and at a price that college students can afford. By that time the "standard" calculus course will have changed considerably, away from the routine calculational skills that occupy much of our class time now, and toward emphases on concepts and on strategies for solving more realistic problems. (The topics you are studying in this book are intended to be consistent with those emphases.)

But low-cost, handheld, symbolic calculators are not with us yet, so we have to compute with the tools that are available. Furthermore, a complicated symbolic expression for a derivative (i.e., a rate of change) is often *not* a practical solution to the problem at hand. If one wants *values* of the derivative, it is more efficient (as we will soon see) to obtain them from a small number of values of the original function than from evaluation of possibly a

Figure 32.1

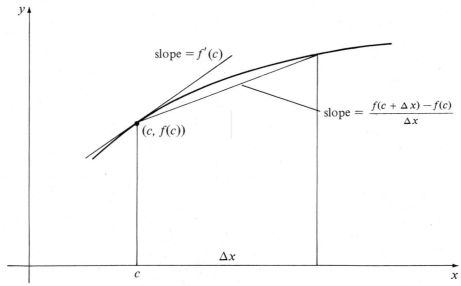

more complicated function. In this chapter we take the first small step toward such a method; the second step appears in Chapter 34.

What's wrong with computing values of the derivative as limits of difference quotients? As Figure 32.1 shows, we can't really expect a difference quotient to be *very* close to the value of the derivative unless Δx is quite small. But when Δx is very small, the numerator is computed as a difference of function values that are very close together, which means cancellation of perhaps several significant digits; and subsequent division by a very small number has the effect of magnifying the loss of significance into possibly a very large error.[2] (It should be noted that *some* difference quotients don't present this problem [e.g., those for which $f(c) = 0$, and no subtraction actually takes place]. In certain other cases there are tricks for writing difference quotients in other forms before computation takes place,[3] but we are seeking a more generally applicable method.)

32.2 A QUADRATIC APPROXIMATION

Our approach to the problem of computing $f'(c)$ at a fixed but arbitrary number c is the same as that employed in Chapter 29[4] for computing an integral: We approximate f in the vicinity of c by a quadratic polynomial $p(x)$, and take $p'(c)$ as our approximation to $f'(c)$.

---

[2]For a more detailed discussion of this problem, see Hamming, pp. 1–5; Forsythe, Sections 1, 2, 3, and 5; or Franta. See also Chapter 14, where a related problem is discussed.

[3]See Hamming, pp. 5–9. See also Chapter 14.

[4]If you have not already studied Chapter 29, please read just the first paragraph of Section 29.1.

For convenience, we first consider the case of $c = 0$. Our result will be phrased in terms of function values and spacing between them (as was the case with Simpson's Rule), and thus may be easily translated to an arbitrary c on the x-axis. Our requirement for the approximating polynomial $p(x)$ is that it should agree with f at $-h$, 0, and h (see Fig. 29.1, where the same situation is considered). This requirement is equivalent to equations (29.2), in which A, B, and C represent the unknown coefficients of p as given by (29.1). Differentiating (29.1) formally, we see that

$$p'(x) = 2Ax + B, \tag{32.2}$$

and hence $p'(0) = B$. Thus it is the middle coefficient of p that we need to determine. From the first and last of equations (29.2), we see that $y_2 - y_0 = 2Bh$, so

$$B = p'(0) = \frac{1}{2h}(y_2 - y_0), \tag{32.3}$$

where $y_2 = f(h)$ and $y_0 = f(-h)$. (This computation is part of the solution of Exercise 29.4.) If we now shift our attention to an interval centered at $x = c$, and replace y_2 by $f(c + h)$ and y_0 by $f(c - h)$, the corresponding approximation to $f'(c)$ becomes

$$p'(c) = \frac{1}{2h}[f(c + h) - f(c - h)]. \tag{32.4}$$

We could have arrived at (32.4) as an approximation to $f'(c)$ in a different way (see Fig. 32.2). The right-hand side is just the slope of the line joining the points $(c - h, f(c - h))$ and $(c + h, f(c + h))$, which should certainly approximate the slope of the tangent line at $(c, f(c))$ if h is small. As a by-product of our derivation by fitting $p(x)$, however, we also discover yet another property of parabolas known to Archimedes: The chord of a parabola (with vertical axis) joining the points $(a, p(a))$ and $(b, p(b))$ is parallel to the tangent line at the midpoint of $[a, b]$ (again, see Fig. 32.2, and also Fig. 28.3).

Figure 32.2

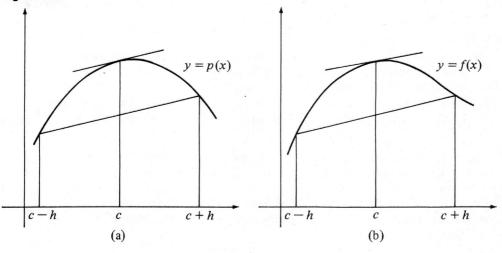

(a) (b)

To use formula (32.4) intelligently, we need some idea of how close $p'(c)$ should be to $f'(c)$ for a given value of h. In the next section we see that

$$|f'(c) - p'(c)| \leq \frac{K}{6} h^2, \tag{32.5}$$

where K is any constant such that $|f'''(x)| \leq K$ for $c - h \leq x \leq c + h$.

This result is of theoretical interest more than it is a computational formula for controlling error. It says, for example, that if $h = 10^{-4}$, then the error should be on the order of 10^{-8}, provided K isn't too large. For most functions, $h = 10^{-4}$ is large enough to ensure that formula (32.4) will not present loss-of-significance numerical problems, even though it is a difference quotient. However, it would be unreasonable to suppose that we would actually control the errors by computing a bound for the third derivative, when the problem we intend to solve is to compute values of the first derivative.

Exercises

1. Prepare a program to tabulate f and its derivative f' over an interval $[a, b]$ in steps of size s. This program should resemble Program 6.1 in form with the following exceptions: (a) We want an additional input variable h to allow experimentation with formula (32.4), and (b) there will need to be an additional item in the output line for the computed value of f'.

2. [O] The quadratic-curve fit can also be used to compute approximations to $f''(c)$. Retrace the steps of our derivation to show that the appropriate formula for the approximation is

$$p''(c) = \frac{1}{h^2} [f(c - h) - 2f(c) + f(c + h)] \tag{32.6}$$

(see Exercise 4 in Chap. 29). This formula may be added to your program if you would like to tabulate f'' along with f and f'.

3. [O] (Discussion questions)
   (a) Formula (32.4) uses only two of the three available items of information about f; $f(c)$ does not appear. Can you think of a reason that the value of f at c has no effect on the computed answer? [Note that $f(c)$ *is* used in (32.6).]
   (b) Why is (32.6) likely to give less satisfactory answers for f'' than (32.4) gives for f'?
   (c) What happens if we try to use the quadratic fit to derive a formula for approximating f'''?

4. [O] Since the error estimate (32.5) cannot be incorporated directly into a differentiation program, the usual way to control the error is to start each computation of $f'(c)$ with $h = 0.01$, say, and then successively halve h until recomputing (32.4) with a smaller h no longer makes a difference (to within some ϵ). One obtains from (32.5) a measure of confidence that this will be a rapidly converging process. Modify your program to include this device.

32.3 HOW GOOD IS THE APPROXIMATION? [O]

In this section we present a proof of the error formula (32.5). For convenience, we consider again the case $c = 0$ and introduce an error function for the polynomial approximation:

$$E(x) = f(x) - p(x), \qquad -h \le x \le h \tag{32.7}$$

(see Fig. 29.1 again). Then

$$E'(x) = f'(x) - p'(x), \tag{32.8}$$

and we want to find a bound for $|E'(0)|$. We assume that f has a continuous third derivative on $[-h, h]$ so that a bound K for f''' exists. Our principal tool in estimating the size of $|E'(0)|$ is *Rolle's Theorem*, which says that if g is a differentiable function such that $g(a) = g(b) = 0$, then somewhere between a and b, g' takes the value 0.

We know that $E(x) = 0$ for $x = -h, 0$, and h (the points at which f and p were required to agree). Another function that has this property is the cubic polynomial

$$P(x) = x(x - h)(x + h) = x^3 - h^2 x. \tag{32.9}$$

For future reference, note that $P'(x) = 3x^2 - h^2$, so

$$P'(0) = -h^2. \tag{32.10}$$

Also note that differentiation twice more gives

$$P'''(x) = 6, \qquad \text{for all } x. \tag{32.11}$$

Both E and P have the general shape illustrated in Figure 32.3. Notice the leveling off (zero slope) at least once in $(-h, 0)$ and once in $(0, h)$—that's Rolle's Theorem. Another function with this same general shape is

$$F(x) = h^2 E(x) + E'(0)P(x). \tag{32.12}$$

Clearly, $F(-h) = F(0) = F(h) = 0$.

WARNING: This is a rabbit-out-of-the-hat trick! The only reason we can offer for introducing F is that it works, in the sense that it leads to our desired estimate, as we will see.

Now, as noted before, Rolle's Theorem tells us that there is a number t_1 in $(-h, 0)$ such that $F'(t_1) = 0$, and also a number t_2 in $(0, h)$ such that $F'(t_2) = 0$. However, F' has at least *one more* root: We have $F'(x) = h^2 E'(x) + E'(0)P'(x)$, so $F'(0) = h^2 E'(0) + E'(0)P'(0) = 0$, by (32.10). Hence the true shape of F is more like Figure 32.4.

Since F' has roots at t_1, 0, and t_2, it too has a shape like Figure 32.3, but with t_1 replacing $-h$ and t_2 replacing h. Hence its derivative, i.e., F'', has at least two roots between t_1 and t_2, and one more application of Rolle's Theorem, this time to F'', tells us that F''' has at least one root, say,

$$F'''(t) = 0, \tag{32.13}$$

for some t in $(-h, h)$. Now let's go back to the definition (32.12) of $F(x)$, and calculate $F'''(x)$, using (32.7) and (32.11):

$$F'''(x) = h^2 E'''(x) + E'(0)P'''(x)$$

$$= h^2 [f'''(x) - p'''(x)] + 6E'(0)$$

$$= h^2 f'''(x) + 6E'(0),$$

Figure 32.3   **Figure 32.4**

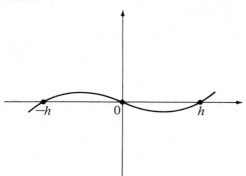   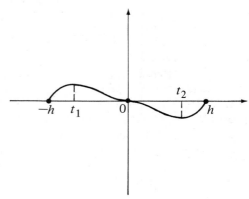

where in the last step we have used the fact that $p'''(x) = 0$ because p is a quadratic polynomial. Now if we substitute $x = t$, use (32.13), and solve for $E'(0)$, we get

$$E'(0) = -\frac{1}{6} h^2 f'''(t). \qquad (32.14)$$

Finally, after taking absolute values, (32.14) and the definition of K imply

$$| E'(0) | \le \frac{K}{6} h^2, \qquad (32.15)$$

which is our desired bound for the difference between $p'(0)$ and $f'(0)$.

References

Coolidge (1951); Grabiner; Hamming; D.A. Smith (1975).

C H A P T E R 3 3

LET'S DIFFERENTIATE [L]

The following exercises call for tabulating various functions and their derivatives, using your program from Chapter 32. In each of the first four cases, $h = 0.0001$ will produce good answers for the derivatives, but you may want to experiment with both larger and smaller values of h to see what the effect will be. Note that use of a smaller value of h does *not* require more computer time, in contrast to integral computations, where a demand for greater accuracy (smaller ϵ) forces smaller step sizes h and more terms in whatever sum is to be evaluated.

In each exercise, you are given a function f and an interval $[a, b]$. Tabulate f and f' at steps of s equal to about $(b - a)/10$.

Exercises

1. $f(x) = x^4 + 2x^3 - 3x^2 + 1$ on $[-3, 2]$. (Program checkout: Check several values of the derivative by hand calculation.)

2. $f(x) = \sin x$ on $[0, 1]$.

NOTE: This exercise can also be used for checking accuracy of answers by doing the companion problem

$$f(x) = \cos x \text{ on } [0, 1].$$

As sine and cosine are, except for sign, derivatives of each other, the f tabulation from each part checks the f' tabulation from the other.

3. $f(x) = (x - 1)/(x + 1)$ on $[0, 5]$.   (Hand checking is possible here.)

4. Use the function f defined by (32.1) on $[1, 6]$.

NOTE: The hardest part of this problem is typing the function definition. Be careful about exponents: For example, ( )$^{2/7}$ should be entered as ( ) $\uparrow$ (2/7). If you omit the parentheses around 2/7, the expression means "square and divide by 7."

5. $f(x) = x \sin (1/x)$ on $[0.001, 0.01]$.

6. $f(x) = x^2 \sin (1/x)$ on $[0.001, 0.01]$.

NOTE: A convenient choice of s for the two preceding exercises is 0.001. In both these cases, $h = 0.0001$ produces answers, but you should not be too easily satisfied with them. Try some smaller values of h to see what happens. In each case, you can take advantage of the capabilities of the computer to do some checking (see the following exercises).

Figure 33.1

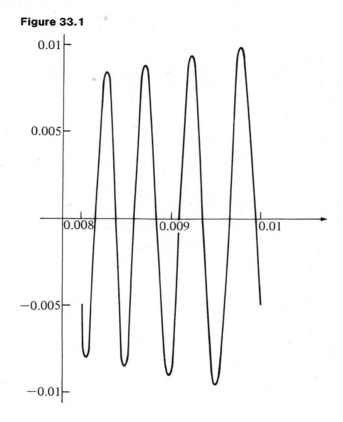

7. The derivative of the function in Exercise 5 may be computed by hand. Check that this derivative is

$$f'(x) = \sin\frac{1}{x} - \frac{1}{x}\cos\frac{1}{x}.$$

This function may be tabulated with Program 6.1 and the tabulation used to check your answers to Exercise 5.

8. Differentiate the function in Exercise 6 by hand, and use the result to check several answers from that exercise.

NOTE: If your answers from Exercises 5 and 6 don't check, rework the exercises with smaller values of h. A partial graph of $f(x) = x \sin(1/x)$ is shown in Figure 33.1. Does this help you see why smaller values of h may be needed? (A similar function was the subject of Sect. 6.4.)

CHAPTER 34

HOW TO REALLY DIFFERENTIATE

34.1 A QUARTIC APPROXIMATION

The parabolic-curve fit used in Chapter 32 for numerical differentiation works reasonably well if the function being differentiated is not too complicated and if h is chosen carefully. However, Exercises 33.5 and 33.6 reveal some inherent limitations in trying to work with only three points on the graph of the given function. Moreover, the differentiation formula (32.4) is so simple that one might suspect that it would not be too difficult to fit a higher-degree polynomial to more points on the curve. The purpose of doing this is to get better results with values of h that are not so small, thereby reducing the likelihood of the loss-of-significance problems hinted at earlier.

In this chapter we derive a differentiation formula using a fourth-degree (quartic) polynomial fitted to five points on the graph of the given function. This choice is convenient because it allows us to use values of f at $-2h$, $-h$, 0, h, and $2h$ [for computing $f'(0)$], and the steps of the derivation are very similar to those in Chapter 32. In particular, the result for approximating $f'(0)$ in terms of five function values and the spacing h is easily translated to an arbitrary center point $x = c$ for purposes of computing $f'(c)$.

The general form for a fourth-degree polynomial is

$$p(x) = Ax^4 + Bx^3 + Cx^2 + Dx + E. \tag{34.1}$$

Let $y_0 = f(-2h)$, $y_1 = f(-h)$, $y_2 = f(0)$, $y_3 = f(h)$, $y_4 = f(2h)$. The requirement that p should agree with f at these five values of x is the following system of equations in the five unknown coefficients A, B, C, D, and E:

$$16Ah^4 - 8Bh^3 + 4Ch^2 - 2Dh + E = y_0$$

$$Ah^4 - Bh^3 + Ch^2 - Dh + E = y_1$$

$$E = y_2 \tag{34.2}$$

$$Ah^4 + Bh^3 + Ch^2 + Dh + E = y_3$$

$$16Ah^4 + 8Bh^3 + 4Ch^2 + 2Dh + E = y_4.$$

Rather than solve these equations completely, we note that all we really need to know is $p'(0)$. Because

$$p'(x) = 4Ax^3 + 3Bx^2 + 2Cx + D, \tag{34.3}$$

we have $p'(0) = D$, so only one of the five coefficients need be determined. Also note that subtracting the first equation from the last and the second from the fourth produces two equations involving only B and D:

$$16Bh^3 + 4Dh = y_4 - y_0$$

$$2Bh^3 + 2Dh = y_3 - y_1. \tag{34.4}$$

We may now eliminate B by subtracting eight times the second equation in (34.4) from the first:

$$-12Dh = y_4 - y_0 - 8(y_3 - y_1).$$

Finally,

$$D = \frac{1}{12h}(y_0 - 8y_1 + 8y_3 - y_4),$$

which gives our approximation $p'(0)$ to $f'(0)$ in terms of h and (four out of five) function values. As in Chapter 32, we now shift the result from $x = 0$ to $x = c$: If $y_0 = f(c - 2h)$, $y_1 = f(c - h)$, $y_2 = f(c)$, $y_3 = f(c + h)$, and $y_4 = f(c + 2h)$, and if a fourth-degree polynomial p is required to agree with f at these points, then,

$$p'(c) = \frac{1}{12h}(y_0 - 8y_1 + 8y_3 - y_4). \tag{34.5}$$

Clearly this approximation to $f'(c)$ is not substantially more difficult to compute than was (32.4), and it is based on a fitted curve that ought to be, in some sense, "twice as good." In

Figure 34.1

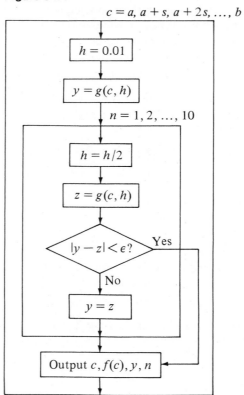

Chapter 36 we see that there is reason to believe that the error is actually proportional to h^4, which means that (34.5) ought to produce (typically) twice as many correct decimal places for a given h as does (32.4). (The error for the latter is proportional to h^2, as we have seen.)

34.2 GETTING THE RIGHT SPACING

In our previous computations of values of f', we chose a spacing h more or less arbitrarily. In practice, this arbitrariness is avoided by starting with a modest value for h (say, $h = 0.01$), and successively halving h until recomputing (34.5) with a smaller h no longer makes a difference, up to a desired number of decimal places. (This approach was suggested earlier in optional Exercise 4 of Chap. 32.) In other words, each derivative value is computed as the limit of a sequence, but note it is a sequence that we can expect to converge very rapidly because of the error estimate hinted at before. The logic of this interval-halving technique is shown in the partial flowchart of Figure 34.1. In that flowchart, $g(c, h)$ refers to the right-hand side of (34.5).

As a practical matter, there needs to be a limit imposed on successive halving of h—our flowchart suggests that no more than ten halving steps be allowed (note indexing of inner loop). This allows h to range down to $0.01/2^{10}$, or slightly less than 10^{-5}, which is adequate for the successful differentiation of any reasonable function and avoids numerical significance problems.

The normal route to the output step in the flowchart is the one resulting from interruption of the inner loop. Note that c represents the current value of x, y the accepted value of $f'(c)$, and n the number of times h was halved in order to get an acceptable y. (If the inner loop is completed, most computer systems will print the value of n as 11, which is the tipoff that the corresponding value of y should be viewed skeptically.)

Exercise

1. Complete the flowchart of Figure 34.1, and write a program to tabulate f and f' over $[a, b]$ with spacing s between tabulated points. The end-test tolerance ϵ for computing f' should be an input variable.

34.3 HIGHER DERIVATIVES [O]

Having insisted that our approximating polynomial p agree with f at five points, we would not be surprised if the second derivative of f could be approximated quite well by p'', and perhaps higher derivatives as well. Looking back at (34.3), we see that

$$p''(x) = 12Ax^2 + 6Bx + 2C,$$

so $p''(0) = 2C$. Thus we need to determine C from the system of equations (34.2). This time we *add* the first and last equations, and also the second and fourth, to eliminate B and D:

$$32Ah^4 + 8Ch^2 + 2E = y_0 + y_4,$$
$$2Ah^4 + 2Ch^2 + 2E = y_1 + y_3.$$

The middle equation of (34.2) says that E is just y_2. Hence

$$32Ah^4 + 8Ch^2 = y_0 + y_4 - 2y_2,$$
$$2Ah^4 + 2Ch^2 = y_1 + y_3 - 2y_2. \tag{34.6}$$

Now A may be eliminated from (34.6) by subtracting 16 times the second equation from the first:

$$-24Ch^2 = y_0 - 16y_1 + 30y_2 - 16y_3 + y_4. \tag{34.7}$$

Finally, we divide through by $-12h^2$ to solve for $p''(0) = 2C$:

$$2C = \frac{1}{12h^2}(-y_0 + 16y_1 - 30y_2 + 16y_3 - y_4).$$

As before, we shift to $x = c$ to get our approximation to $f''(c)$:

$$p''(c) = \frac{1}{12h^2}(-y_0 + 16y_1 - 30y_2 + 16y_3 - y_4). \tag{34.8}$$

Exercises

2. Modify your program from Exercise 1 to include output of $f''(x)$, computed from (34.8) for the last value of h only. (Note that no error estimate or control is being suggested here.)

3. Derive an approximation formula for $f'''(x)$ from the quartic-curve fit.

HINT: Solve (34.4) for B.

4. Modify the discussion questions of Exercise 3 of Chapter 32 to make them appropriate to this chapter. Discuss.

CHAPTER 35

LET'S REALLY DIFFERENTIATE [L]

The problems in this laboratory session are similar to those in Chapter 33, but you are to use your program from Chapter 34. Except as noted, start with $\epsilon = 0.0005$, but feel free to experiment with smaller values of ϵ as well. Each exercise specifies a function to be differentiated and an interval over which the tabulation is to be carried out.

Exercises

1. $f(x) = x^4 + 2x^3 - 3x^2 + 1$ on $[-3, 2]$. (Program checkout. What sort of accuracy do you expect to get for this function?)

2. $f(x) = \tan x$ on $[-1, 1]$.

NOTE: If your system does not have a built-in tangent function, use $SIN(X)/COS(X)$.

3. $f(x) = (\sin x - \cos x)/\sqrt{1 + x^2}$ on $[-1, 1]$.

4. $f(x)$ as defined by (32.1) on $[0.5, 1.5]$. (See Exercise 4 of Chap. 33 for an important message before proceeding.)

5. $f(x) = [\cos (x^2 - 1) + x^2]^{1/2}/(3x^2 + 17)^{1/3}$ on $[3, 5]$. (See Exercise 4 of Chap. 33.)

Before you attempt the last two exercises, a small change in your program is needed. Each of these exercises involves a tabulation starting at 0.001 for a function that is discontinuous at $x = 0$. Change your starting value for h from 0.01 to 0.0001 so that $c - 2h$ and $c - h$ will never be negative or zero. For Exercise 6 only, start with $\epsilon = 0.05$ before trying any smaller values (see Fig. 33.1).

6. $f(x) = x \sin (1/x)$ on $[0.001, 0.01]$.

7. $f(x) = x^2 \sin (1/x)$ on $[0.001, 0.01]$.

CHAPTER 36

ERROR ANALYSIS VIA ROLLE'S THEOREM AND LEGERDEMAIN [O]

In Section 32.3 we analyzed the error in the quadratic differentiation formula by using Rolle's Theorem and a rabbit-out-of-the-hat trick. Chapter 34 claimed that the quartic differentiation formula is "twice as good," and perhaps you have seen some evidence of that in your last laboratory session. Here we demonstrate exactly what that means; the development is very similar to Section 32.3, except that there are *two more of everything* (derivatives, roots, etc.).

To clarify the role of Rolle's Theorem, we begin by sharpening this basic tool. Figure 36.1 shows a function that has four roots, and we suppose that it is at least thrice differentiable. Then Rolle's Theorem says that f' has a root between r_1 and r_2, another between r_2 and r_3,

Figure 36.1

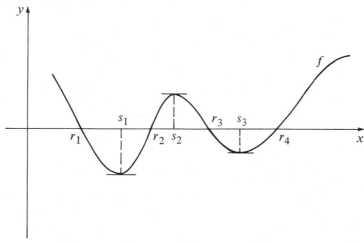

and still another between r_3 and r_4—i.e., f' has at least three different roots, say, s_1, s_2, s_3. But then we may apply Rolle's Theorem to f' to conclude that its derivative (f'') has roots between s_1 and s_2 and between s_2 and s_3. Since f'' has at least two roots, yet another application of Rolle's Theorem leads us to conclude that f''' has at least one root between r_1 and r_4. That is, the existence of four roots of f implies the existence of at least one root of f''', by multiple application of Rolle's Theorem. The general case of this result may be summed up as follows:

Rolle's Theorem Extended   If, on some interval, the function f is k times differentiable and has at least $k + 1$ roots, then its kth derivative has at least one root.

We turn now to the situation considered in Chapter 34: p is a quartic polynomial that agrees with the given function f at $-2h$, $-h$, 0, h, and $2h$, and $f'(0)$ is approximated by $p'(0)$; so our task is to estimate $|f'(0) - p'(0)|$. Following the "two more of everything" principle, we assume that f has a continuous *fifth* derivative on $[-2h, 2h]$ and that K is a number such that $|f^{(5)}(x)| \le K$ for $-2h \le x \le 2h$.

Define the error function

$$E(x) = f(x) - p(x), \qquad -2h \le x \le 2h. \tag{36.1}$$

We know that $E(x) = 0$ at $x = 0$, $\pm h$, $\pm 2h$. Another function that has this property is the polynomial

$$P(x) = x(x - h)(x + h)(x - 2h)(x + 2h) = x^5 - 5h^2x^3 + 4h^4x. \tag{36.2}$$

Note that

$$P'(0) = 4h^4, \tag{36.3}$$

and

$$P^{(5)}(x) = 120, \qquad \text{for all } x. \tag{36.4}$$

Here comes the rabbit: For no apparent reason,[1] except that it works, we introduce a third function F sharing the five roots of E and P:

$$F(x) = 4h^4 E(x) - E'(0)P(x). \tag{36.5}$$

Since F has the five roots 0, $\pm h$, $\pm 2h$, its derivative F' has at least four roots between consecutive pairs of these numbers, by the usual version of Rolle's Theorem. But F' also has the fifth root zero:

$$F'(0) = 4h^4 E'(0) - E'(0)P'(0) = 0,$$

by (36.3). Hence the fourth derivative of F', i.e., $F^{(5)}$, has a root in $(-2h, 2h)$, by the extended version of Rolle's Theorem—i.e., $F^{(5)}(t) = 0$ for some t.

Now let's compute that fifth derivative:

$$\begin{aligned}
F^{(5)}(x) &= 4h^4 E^{(5)}(x) - E'(0)P^{(5)}(x) \\
&= 4h^4 [f^{(5)}(x) - p^{(5)}(x)] - 120E'(0) \\
&= 4h^4 f^{(5)}(x) - 120E'(0),
\end{aligned}$$

using (36.4) and the fact that $p^{(5)}(x) = 0$, since p is quartic. Substituting $x = t$, we have

$$0 = 4h^4 f^{(5)}(t) - 120E'(0),$$

or

$$E'(0) = \frac{1}{30} h^4 f^{(5)}(t).$$

Finally, taking absolute values, and using our bound K for $f^{(5)}$, we have

$$|E'(0)| \le \frac{K}{30} h^4. \tag{36.6}$$

Since $E'(0) = f'(0) - p'(0)$, we have the desired estimate for the error in formula (34.5). Comparison with (32.13) shows why this formula is "twice as good" as the quadratic approximation (but note that the constants "K" in the two estimates are not the same).

Exercise

Generalize the preceding technique to establish the following theorem:[2] Let f be a function with a continuous $(2k + 1)$-th derivative on the interval $[-kh, kh]$, where k is a positive integer and h is a positive real number. Let K be any bound for $|f^{(2k+1)}(x)|$ on the same interval. Let p be a polynomial of degree $2k$ that agrees with f at the points $0, \pm h, \pm 2h, \ldots, \pm kh$. Then

$$|f'(0) - p'(0)| \le \frac{K(k!)^2 h^{2k}}{(2k + 1)!}.$$

---

[1]Although the reason may not be apparent, there is one, of course. Any function of the form $F(x) = c_1 E(x) + c_2 P(x)$ will share the common roots of E and P. The object is to choose c_1 and c_2 so that $F'(0) = 0$, which requires only that $c_1 E'(0) = -c_2 P'(0)$.

[2]See D. A. Smith (1975).

C H A P T E R   3 7

A DO-IT-YOURSELF KIT
FOR SINES AND COSINES

37.1 APPROXIMATIONS BY POLYNOMIALS

In earlier chapters, particularly in laboratory sessions, we have made frequent use of built-in sine and cosine functions as available features of the computer system. Now that you have studied the calculus of these trigonometric functions, perhaps you have wondered how the computer (basically a big *arithmetic* machine) knows how to evaluate these functions. One possible answer (not necessarily the right one for your particular computer) lies in a clever sort of polynomial approximation to nonpolynomial functions that is the subject of this and several succeeding chapters.

The last 11 chapters (numerical integration and differentiation) have made use of polynomial approximations to functions that were obtained by making the function and the approximating polynomial agree at certain specified points. To use such approximations effectively, one must be able to evaluate the given function at an arbitrary point in its domain. Suppose we *don't* know how to do that—e.g., suppose our computer system happened *not* to be equipped with built-in trig functions and we needed to evaluate sin x for some unfamiliar angle x. If we can find a polynomial that agrees closely with our given function *throughout* some interval, and if the determination of this polynomial does not require computation of function values, then we are all set, since our arithmetic machine can easily compute values of polynomials. Our basic tool for obtaining polynomial approximations to sin x and cos x is a property of the definite integral for continuous functions (also used earlier in optional Chaps. 28 and 31):

$$\text{If} \quad g_1(x) \le g_2(x) \quad \text{for } a \le x \le b,$$
$$\text{then} \quad \int_a^b g_1(x)\, dx \le \int_a^b g_2(x)\, dx. \tag{37.1}$$

(Recall that this is a very easy consequence of the definition of the integral as a limit of sums.) We also need the following facts:

$$\int_0^x \cos t\, dt = \sin t \,\big|_0^x = \sin x. \tag{37.2}$$

and

$$\int_0^x \sin t\, dt = -\cos t \,\big|_0^x = 1 - \cos x. \tag{37.3}$$

Our starting point is the obvious inequality

$$\cos t \le 1 \quad \text{(for all } t\text{)}. \tag{37.4}$$

If we integrate both sides of (37.4) from 0 to x, using (37.1) and (37.2), we obtain

$$\sin x \le x, \qquad x \ge 0. \tag{37.5}$$

NOTE: (37.1) requires that the upper limit of integration be greater than or equal to the lower limit. If this is not the case, the resulting sign change reverses all inequalities. To avoid this complication, we limit our attention to nonnegative values of x for the time being.

Now we rewrite (37.5) as $\sin t \le t$, and integrate again from 0 to x:

$$1 - \cos x \le \frac{x^2}{2},$$

or

$$1 - \frac{x^2}{2} \le \cos x, \qquad x \ge 0. \tag{37.6}$$

The next step is to replace x by t in (37.6) and integrate again from 0 to x:

$$x - \frac{x^3}{6} \le \sin x, \qquad x \ge 0. \tag{37.7}$$

Now the pair (37.4) and (37.6) gives us the following bounds for the cosine function:

$$1 - \frac{x^2}{2} \le \cos x \le 1, \qquad x \ge 0, \tag{37.8}$$

and, similarly, the pair (37.5) and (37.7) gives bounds for the sine function:

$$x - \frac{x^3}{6} \le \sin x \le x, \qquad x \ge 0. \tag{37.9}$$

We proceed from (37.9) as before: Change x to t, and integrate all three terms of the inequality from 0 to x:

$$\frac{x^2}{2} - \frac{x^4}{24} \le 1 - \cos x \le \frac{x^2}{2},$$

or

$$1 - \frac{x^2}{2} \le \cos x \le 1 - \frac{x^2}{2} + \frac{x^4}{24}, \qquad x \ge 0. \tag{37.10}$$

The same process may be repeated as many times as we wish, each time producing new polynomial bounds for $\sin x$ or $\cos x$ involving more terms and higher degrees than before. The next two sets of bounds are

$$x - \frac{x^3}{6} \le \sin x \le x - \frac{x^3}{6} + \frac{x^5}{120}, \qquad x \ge 0, \tag{37.11}$$

and

$$1 - \frac{x^2}{2} + \frac{x^4}{24} - \frac{x^6}{720} \le \cos x \le 1 - \frac{x^2}{2} + \frac{x^4}{24}, \qquad x \ge 0. \tag{37.12}$$

(Check these carefully before you proceed.)

In order to state the general result, we introduce some abbreviations for the polynomials that are emerging:

$$S_{2n+1}(x) = x - \frac{x^3}{3!} + \frac{x^5}{5!} - \cdots + (-1)^n \frac{x^{2n+1}}{(2n+1)!} \qquad (37.13)$$

and

$$C_{2n}(x) = 1 - \frac{x^2}{2!} + \frac{x^4}{4!} - \cdots + (-1)^n \frac{x^{2n}}{(2n)!}. \qquad (37.14)$$

In both cases, n is allowed to take the values 0, 1, 2, 3, ... and the subscript of the polynomial name is the degree of the polynomial.

Our last inequalities (37.11) and (37.12) may be shortened to

$$S_3(x) \leq \sin x \leq S_5(x)$$

and

$$C_6(x) \leq \cos x \leq C_4(x).$$

If we were to proceed with the algorithm of changing the variable and integrating from 0 to x, we would find

$$S_7(x) \leq \sin x \leq S_5(x)$$

and

$$C_6(x) \leq \cos x \leq C_8(x)$$

and so on. The point is that, for each fixed x (at least if $x \geq 0$), $\sin x$ lies *between* any two consecutive terms of the sequence $S_1(x), S_3(x), S_5(x), \ldots$. Similarly, $\cos x$ lies *between* any two consecutive terms of the sequence $C_0(x), C_2(x), C_4(x), \ldots$. What makes these facts *useful* is that consecutive terms of these sequences get closer and closer to each other, so that anything caught between every pair of consecutive terms is approximated more and more closely by the terms themselves as we go farther out in the sequence. In the language of Chapter 7, we are saying that

$$\sin x = \lim_{n \to \infty} S_{2n+1}(x) \qquad (37.15)$$

and

$$\cos x = \lim_{n \to \infty} C_{2n}(x) \qquad (37.16)$$

for each $x \geq 0$. At this point we might observe that the restriction to $x \geq 0$ is unnecessary. As noted earlier, if $x < 0$, each integration step from 0 to x changes a sign and reverses all inequalities; but the conclusion is still the same: $\sin x$ is *between* $S_{2n+1}(x)$ and $S_{2n+3}(x)$, and $\cos x$ is *between* $C_{2n}(x)$ and $C_{2n+2}(x)$, for every n.

Figures 37.1 and 37.2 show the approximating polynomials for $\sin x$ and $\cos x$ up through degree 19. (Note that the subscript notation in the figures is different from that used previously.)

Figure 37.1

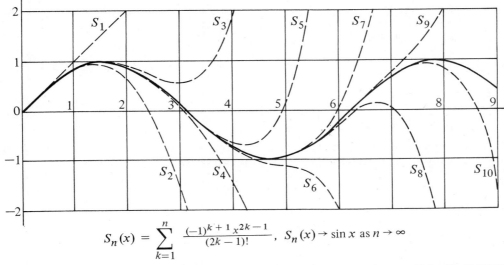

$$S_n(x) = \sum_{k=1}^{n} \frac{(-1)^{k+1} x^{2k-1}}{(2k-1)!}, \quad S_n(x) \to \sin x \text{ as } n \to \infty$$

From Kammerer, H.M. "Sine and Cosine Approximation Curves." *American Mathematical Monthly* 43 (1936), 293–294.

Figure 37.2

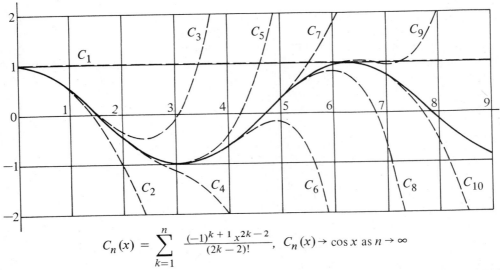

$$C_n(x) = \sum_{k=1}^{n} \frac{(-1)^{k+1} x^{2k-2}}{(2k-2)!}, \quad C_n(x) \to \cos x \text{ as } n \to \infty$$

From Kammerer, H.M. "Sine and Cosine Approximation Curves." *American Mathematical Monthly* 43 (1936), 293–294.

37.2 WHERE DOES IT ALL STOP?

Let's see why consecutive terms of the two sequences actually get close to each other. The definitions (37.13) and (37.14) may be rewritten in the following, more useful, forms:

$$S_{2n+1}(x) = S_{2n-1}(x) + (-1)^n \frac{x^{2n+1}}{(2n+1)!} \qquad (37.17)$$

and

$$C_{2n}(x) = C_{2n-2}(x) + (-1)^n \frac{x^{2n}}{(2n)!}. \qquad (37.18)$$

These recursive definitions, together with the starting values $S_1(x) = x$, $C_0(x) = 1$, tell us how to actually compute consecutive terms of the two sequences. They also tell us that

$$|S_{2n+1}(x) - S_{2n-1}(x)| = \frac{|x|^{2n+1}}{(2n+1)!} \qquad (37.19)$$

and

$$|C_{2n}(x) - C_{2n-2}(x)| = \frac{|x|^{2n}}{(2n)!}. \qquad (37.20)$$

The right-hand members of (37.19) and (37.20) are, respectively, the odd-numbered and even-numbered terms of the sequence $\{r^k/k!\}$, where $r = |x|$. Back in Chapter 8 you had an opportunity to observe some empirical evidence that the limit of this sequence is zero for any real number r; the time has come to establish that fact.

Theorem   If r is any real number, then

$$\lim_{k \to \infty} \frac{r^k}{k!} = 0.$$

Proof   [O] By Property 9.3, it is sufficient to show that $\lim_{k \to \infty} |r|^k/k! = 0$, so we may as well assume that $r > 0$. (The case of $r = 0$ is trivial.) Recall that only the "tail" of a sequence matters (Property 9.1), so we may start as far out in the sequence as we like. Let M be any fixed integer such that $M > r$, and consider only terms of the sequence for which $k > M$. Then

$$\frac{r^k}{k!} = \frac{r \cdot r \cdot r \cdots r \cdot r \cdots r}{1 \cdot 2 \cdot 3 \cdots M(M+1) \cdots k}$$

$$= \frac{r^M}{M!} \cdot \underbrace{\frac{r}{M+1} \cdot \frac{r}{M+2} \cdots \frac{r}{k}}_{k - M \text{ factors}}$$

$$< \frac{r^M}{M!} \left(\frac{r}{M}\right)^{k-M}$$

$$= \frac{M^M}{M!} \left(\frac{r}{M}\right)^k.$$

Hence

$$0 \le \lim_{k \to \infty} \frac{r^k}{k!} \le \lim_{k \to \infty} \frac{M^M}{M!} \left(\frac{r}{M} \right)^k$$
$$= \frac{M^M}{M!} \lim_{k \to \infty} \left(\frac{r}{M} \right)^k = 0,$$

by Property 9.5(c) and Theorem 11.1. (Note that $r/M < 1$.) This completes the proof, the essence of which may be summarized as follows: For sufficiently large values of k, $r^k/k!$ will be "sandwiched" between 0 and a constant multiple of a corresponding term of a geometric progression with ratio less than 1.

An immediate consequence of this theorem is that the right-hand members of (37.19) and (37.20) go to zero as n becomes large (for each fixed value of x). Then the limit formula (37.15) follows from the Pinching Theorem (Property 9.7) and the fact that $\sin x$ lies between $S_{2n-1}(x)$ and $S_{2n+1}(x)$. Similarly, we get (37.16) from the corresponding fact for $\cos x$. But (37.19) and (37.20) contain even more important information. Since $\sin x$ is between $S_{2n-1}(x)$ and $S_{2n+1}(x)$, we conclude from (37.19) that

$$| \sin x - S_{2n-1}(x) | \le \frac{| x |^{2n+1}}{(2n + 1)!}, \qquad (37.21)$$

i.e., we have a bound on the error at any stage of the approximation sequence given by the absolute value of the next term to be added. This gives an obvious way to decide how to stop the process: Compute a new term of the form $(-1)^n x^{2n+1}/(2n + 1)!$; if this term is small enough in absolute value, take $S_{2n-1}(x)$ as the approximation to $\sin x$; if not, add on the new term and repeat. Similarly, we have

$$| \cos x - C_{2n-2}(x) | \le \frac{| x |^n}{(2n)!}, \qquad (37.22)$$

so an exactly analogous algorithm permits the approximation of $\cos x$ by values of polynomials.

Finally, observe that the computation of a "new term" of the sum for $\sin x$ can be streamlined by noting that

$$\frac{(-1)^n x^{2n+1}}{(2n + 1)!} = \frac{(-1)x^2}{(2n + 1)(2n)} \cdot \frac{(-1)^{n-1} x^{2n-1}}{(2n - 1)!}$$
$$= \frac{-x^2}{(2n + 1)(2n)} \cdot (\text{last previous term})$$

Thus if we keep track of the last term added, then we get the next term by multiplying by $-x^2$ and dividing by $(2n + 1)(2n)$.

The essential features of the algorithm are summarized in the partial flowchart shown in Figure 37.3.

Figure 37.3

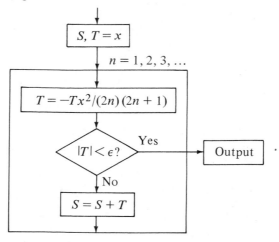

Exercises

1. Complete the flowchart, and write a program that has x and ϵ as input variables and $\sin x$ (to within ϵ) and the degree of the approximating polynomial as output. You might also include $\text{SIN}(X)$ in your output list (using the built-in sine function) to get correct answers for comparison.

2. [O] Repeat Exercise 1 for the cosine function.

37.3 SINES OF LARGE NUMBERS [O]

The form of the error estimates (37.21) and (37.22) makes it clear that the approximating polynomials work better when x is small than when it is large. Indeed, for x quite far from 0, n may have to be very large before $S_{2n-1}(x)$ will really be close to $\sin x$. This should be no surprise, because $|\sin x| \le 1$ for all x, whereas, for any particular polynomial $p(x)$, $|p(x)|$ must become large as $x \to \pm\infty$. Thus the algorithm previously derived may be very impractical if we wanted to know, say, $\sin(2368)$ (see Fig. 37.1 again).

On the other hand, every value of $\sin x$ occurs for x between $-\pi/2$ and $\pi/2$, and the identities

$$\sin(x + 2k\pi) = \sin x \tag{37.23}$$

and

$$\sin(\pi - x) = \sin x \tag{37.24}$$

may be used to replace any x by one in $[-\pi/2, \pi/2]$ having the same sine. Any bound for the error $|1 - S_{2n-1}(\pi/2)|$ will also bound the error $|\sin x - S_{2n-1}(x)|$ for any smaller x (in absolute value); so a single choice of n may be made to ensure a certain accuracy at the endpoints, and then $S_{2n-1}(x)$ will approximate $\sin x$ with at least that accuracy throughout the interval $[-\pi/2, \pi/2]$. For example, as you may verify in the next laboratory session, $|1 - S_9(\pi/2)| < 0.000004$, so $S_9(x)$ agrees with $\sin x$ to at least five places throughout the interval. This is the basis for a general-purpose sine program that does not require the user to limit the size of x or to specify the accuracy requirements, and does not use polynomials of enormously large degree to get the sine of a large number.

Exercises

3. Make a flowchart for, and program, a sine-function routine that replaces the input x by a number y between $-\pi/2$ and $\pi/2$ having the same sine [use (37.23) and (37.24)], and then calculate $S_9(y)$.

NOTE: The recursive method of evaluating S_9 as in Exercise 1 is useful here also, but the loop will not require a special end test.

4. The same approach might be used for computing cosines, but note that, to get all values of the cosine function, one must consider an interval such as $[0, \pi]$. This involves much greater errors (for given n) at the right-hand endpoint, and therefore a much higher degree to achieve desired accuracy. Can you think of a way to avoid this difficulty?

37.4 PREVIEW OF COMING ATTRACTIONS [R]

Several of our early examples of sequences involved terms of the form

$$S_n = a_1 + a_2 + \cdots + a_n = S_{n-1} + a_n$$

(see Sects. 7.1 and 8.2). When $S_n = \Sigma_{k=1}^{n} a_k$, and $\lim_{n \to \infty} S_n$ exists, we write

$$\lim_{n \to \infty} S_n = \sum_{k=1}^{\infty} a_k = a_1 + a_2 + a_3 + \cdots.$$

The resulting "sum" of infinitely many terms is called an *infinite series*. Note that such a series has meaning only as a limit of a sequence, namely, the sequence of "partial sums" $S_1, S_2, S_3, \ldots.$

Using infinite-series notation, we may combine (37.13) and (37.15) to write

$$\sin x = \sum_{k=0}^{\infty} \frac{(-1)^k x^{2k+1}}{(2k+1)!} = x - \frac{x^3}{3!} + \frac{x^5}{5!} - \cdots. \tag{37.25}$$

Similarly, (37.14) and (37.16) may be abbreviated as

$$\cos x = \sum_{k=0}^{\infty} \frac{(-1)^k x^{2k}}{(2k)!} = 1 - \frac{x^2}{2!} + \frac{x^4}{4!} - \cdots. \tag{37.26}$$

Each of these expressions is called a *power series* for the corresponding function because the value of each function at x is expressed as a series involving powers of x. Keep in mind that the computational meaning of a power series for $\sin x$, say, is that one computes terms of the sum on the right in (37.25) until the error estimate (37.21) says that all the remaining terms may safely be ignored. Later in your calculus course, you devote a good deal of attention to the study of convergence of infinite series and to power-series representations of functions. In this and several succeeding chapters, it is more useful to think of sequences of approximating polynomials. All convergence questions that arise are settled by means of error estimates that are similar to (37.21) and (37.22).

37.5 HISTORICAL NOTE [R]

The subject of trigonometry has its origins in the ancient study of astronomy. However, the trigonometric functions as we know them are a relatively recent invention, not really standardized until about 200 years ago. The ancients expressed trigonometric relationships in terms of the ratio of a chord of a circle to the angle it subtended, or, equivalently, to the length of the corresponding arc of the circle. If you think about the usual circle diagram used to define sine, cosine, tangent, and so on, you realize that the sine of an angle expresses the relationship between a half-chord of the circle and half of the angle subtended by the whole chord. The discovery of the usefulness of half-chords is perhaps the most significant contribution of fifth-century Hindu mathematics.

The earliest known tabulation of lengths of chords (the obvious predecessor of sine tables) was done by the astronomer Hipparchus of Nicaea (ca. 180–ca. 125 B.C.), but his method is not known, nor have the tables survived. The oldest surviving table is the Table of Chords constructed by the mathematician-astronomer Ptolemy of Alexandria in the second century A.D. It is not known how much Ptolemy relied on the work of Hipparchus, but he developed the equivalents of the formulas we know today for sines and cosines of sums and differences of angles and of half-angles, and used these to first find sin $\frac{1}{4}°$, and then to build up a table of sines from 0° to 90° in steps of $\frac{1}{4}°$. Actually, he found the length of the side of a regular polygon of 720 sides inscribed in a circle of radius 60, which was also the basis for his four-place approximation to π (see Chap. 12).

The first table of half-chords (essentially a table of sines if each entry is divided by the radius of the circle) was developed in India in the fifth century. The table contains 24 entries for angles from $(3\frac{3}{4})°$ to 90° in steps of $(3\frac{3}{4})°$. This angle is small enough for its sine to be approximately the radian measure of the angle, and that is essentially how the first entry was computed. The successive entries were then computed by a recursion formula involving all previously computed entries. For the details of this method and of Ptolemy's, see Boyer (1968), Chapters X and XII.

37.6 ETYMOLOGICAL NOTE [R]

The word *sine* comes directly from the Latin *sinus*, which has the various meanings *bay, curve, fold, hollow,* or *pocket* (all as nouns). It might be tempting to seize on the meaning "curve," or to associate one arch of the sine curve with the shape of a bay. But remember that the use of sines as chords or half-chords long predates the graphical representation of $y = \sin x$. In fact, the association of the Latin word with the notion of chord of a circle was simply a mistake. The Sanskrit word used by the Hindus for their half-chords was *jiva,* the literal meaning of which was "bowstring," and hence also the chord of an arc. When the Arab mathematicians of the ninth century (see Chap. 3) assimilated, developed, and disseminated the Hindu mathematical works, they adapted the Sanskrit word to an Arabic word, *jiba*. In the twelfth century, when it became important to Europeans to acquire the fruits of the Arab culture, a cosmopolitan group of translators assembled in Spain, where the Arabic books were readily available and the project was officially sponsored by the Church, to translate these works into Latin. One of the most notable of these translators, the Englishman Robert of Chester, was apparently responsible for confusing *jiba* with *jaib,* the Arabic word for bay or inlet; hence his choice of the Latin *sinus.*

Television commercials have made us all painfully aware of the modern use of the word sinus, a legitimate derivation from the Latin meanings "hollow" and "pocket." If you followed the travels of the Apollo astronauts, you may also have noted the use of Latin in naming various sealike and baylike features of the moon, such as Sinus Medii. But sines, unfortunately, have nothing to do with bays.

References

Boyer (1968); Heineman; Higgins; Kammerer; Kuller.

C H A P T E R   3 8

USING THE KIT [L]

Exercises

1. Try out your program for computing sin x with several values of x between 0 and 10 and ϵ small enough to ensure five or six correct decimal places in the answers. Corresponding values for the built-in function SIN may be used for checking purposes.

2. [O] If you have written a cosine program also, check it out in the same way.

3. [O] If you have written a program for Exercise 3 in Chapter 37, you will need an accurate value of π as a constant in the program: Try 3.1415926535898. (If your computer does not allow constants that long, shorten the approximation as appropriate.) Before you can use your program, you need to know that S_9 is sufficiently accurate over the interval $[-\pi/2, \pi/2]$. Use your calculator, a direct inquiry, or a short program to compute $x^{11}/11!$, where $x = \pi/2$ (see Sect. 8.3). If the answer is less than 0.000004, as asserted in Section 37.3, then S_9 will produce answers that are correct to five decimal places [see formula (37.21)].

4. [O] Check out your program from Exercise 3 in Chapter 37 with several different values of x, including some not close to 0. Compare with values of SIN.

5. Modify your program from either Exercise 1 or Exercise 3 of Chapter 37 to produce a table of sines correct to five decimal places over the interval $[0, 2\pi]$, in steps of 0.25. You will need to delete the input statement, set the value of ϵ (if you are using Exercise 1 of Chap. 37), perhaps modify your output statement, and then enclose the whole program in a loop indexed for the appropriate values of x. Then execute the modified program. How does it look? It may not be as pretty as a textbook table, but you did it all by yourself.

6. Use techniques developed earlier in the course to explore the function

$$f(x) = \frac{\sin x}{x}, \qquad x > 0.$$

Determine locations of the first few maxima, minima, and inflection points. (Compute derivatives by hand and roots by computer, using the built-in trig functions.) Compare these with the corresponding points on the sine curve. From your calculus course you should know the limit of $f(x)$ as $x \to 0$. Use all the information obtained to sketch an accurate graph of f on graph paper.

7. [O] Find $\int_0^\pi [(\sin x)/x]\, dx$.

HINT: This can be done by numerical integration, but you will have to figure out what to do at the left-hand endpoint of the interval.

8. [O] Obtain polynomial approximations and error estimates for $f(x) = (\sin x)/x$, starting from the approximations $S_{2n+1}(x)$ for $\sin x$. Reconsider Exercises 6 and 7 in terms of these polynomial approximations.

CHAPTER 39

THE ARCTANGENT FUNCTION AND π

39.1 AN ALGORITHM

Now that we have demonstrated in Chapter 35 how to compute values of trigonometric functions by polynomial approximation, the obvious next step is to do the same for inverse trigonometric functions. Since all six inverse trig functions can be obtained from any one of them, it is convenient to restrict our attention to the arctangent, which is likely to be the one built into your computer system.

One way to compute the arctangent function is to use the Fundamental Theorem of

Calculus and take advantage of the fact that its derivative is a rational function:

$$\arctan x = \int_0^x \frac{dt}{1 + t^2}. \tag{39.1}$$

For any given value of x, the integral on the right may be evaluated by Simpson's Rule (see Chap. 29). However, this is a rather inefficient procedure, and it turns out to be rather easy to derive a polynomial approximation to $\arctan x$ from (39.1).

The first step is to divide $1 + t^2$ into 1 to get a quotient and remainder:

$$\frac{1}{1 + t^2} = 1 - t^2 + t^4 - \cdots + (-1)^n t^{2n} + \frac{(-1)^{n+1} t^{2n+2}}{1 + t^2}, \tag{39.2}$$

for any $n = 0, 1, 2, 3, \ldots$. You may check this by the high school method of long division of polynomials (an unusual application thereof), but note that the process never stops. For *each* n, you get a remainder of the form $(-1)^{n+1} t^{2n+2}$, which is where the last term in (39.2) comes from. ([O] Alternatively, you may refer to the proof of Theorem 11.2, where it is shown that

$$1 + r + r^2 + \cdots + r^n = \frac{1 - r^{n+1}}{1 - r}.$$

If we set $r = -t^2$, this reads

$$1 - t^2 + t^4 - \cdots + (-1)^n t^{2n} = \frac{1 - (-1)^{n+1} t^{2n+2}}{1 + t^2}$$

$$= \frac{1}{1 + t^2} - \frac{(-1)^{n+1} t^{2n+2}}{1 + t^2},$$

which is equivalent to (39.2).)

Now each term of (39.2) may be integrated from 0 to x to produce a formula for $\arctan x$, according to (39.1):

$$\arctan x = x - \frac{x^3}{3} + \frac{x^5}{5} - \cdots + (-1)^{n+1} \int_0^x \frac{t^{2n+2}}{1 + t^2} dt, \tag{39.3}$$

for each $n = 0, 1, 2, \ldots$. Let's abbreviate the polynomial part of the right-hand side of (39.3):

$$A_{2n+1}(x) = x - \frac{x^3}{3} + \frac{x^5}{5} - \cdots + (-1)^n \frac{x^{2n+1}}{2n + 1}. \tag{39.4}$$

This will be our polynomial approximation of degree $2n + 1$ to $\arctan x$, and (39.3) gives a precise description of the error:

$$| \arctan x - A_{2n+1}(x) | = \left| \int_0^x \frac{t^{2n+2}}{1 + t^2} dt \right|. \tag{39.5}$$

All of that is not too exciting, of course, unless the error happens to be *small* or, at worst, gets smaller as n gets larger. Hence we have to find a reasonable bound for the right-hand side of (39.5).

First, note that

$$\left| \int_0^x \frac{t^{2n+2}}{1 + t^2} \, dt \right| = \int_0^{|x|} \frac{t^{2n+2}}{1 + t^2} \, dt,$$

since the function being integrated takes only *positive* values. Now, since $1 + t^2$ is always ≥ 1, we have

$$\frac{t^{2n+2}}{1 + t^2} \leq t^{2n+2}, \qquad \text{for all } t,$$

and hence, by (37.1),

$$\int_0^{|x|} \frac{t^{2n+2}}{1 + t^2} \, dt \leq \int_0^{|x|} t^{2n+2} \, dt = \frac{|x|^{2n+3}}{2n + 3}.$$

Thus we have our error estimate in terms of something we can compute:

$$| \arctan x - A_{2n+1}(x) | \leq \frac{|x|^{2n+3}}{2n + 3}. \tag{39.6}$$

Note that the definition of the polynomials (39.4) can be written recursively as

$$A_{2n+3}(x) = A_{2n+1}(x) + \frac{(-1)^{n+1} x^{2n+3}}{2n + 3}, \tag{39.7}$$

with the starting value $A_1(x) = x$. That is, our bound for the error in $A_{2n+1}(x)$ is the absolute value of the next term to be added on if we were to compute $A_{2n+3}(x)$, a result exactly like that obtained in Chapter 37 for the sine and cosine functions.

However, there is an important difference between (39.6) and the corresponding results in Chapter 37: There is no *factorial* in the denominator this time, and $| x |^{2n+3}/(2n + 3)$ does *not* always go to zero as n becomes large. In particular, if $| x | > 1$, the limit is not zero (it's infinite), but if $| x | \leq 1$, then

$$\lim_{n \to \infty} \frac{| x |^{2n+3}}{2n + 3} \leq \lim_{n \to \infty} \frac{1}{2n + 3} \leq \lim_{n \to \infty} \frac{1}{n} = 0,$$

so

$$\lim_{n \to \infty} A_{2n+1}(x) = \arctan x, \qquad \text{provided } | x | \leq 1. \tag{39.8}$$

Exercise

1. Diagram and program the algorithm described in this section for approximating arctan x to within ϵ, when $-1 \leq x \leq 1$: Compute successively $A_1(x) = x, A_3(x), A_5(x), \ldots$ until the next term to be added on is less than ϵ in absolute value.

39.2 COMPUTATION OF π

Using the "infinite-series" notation introduced in Chapter 37, we may write formula (39.8) as

$$\arctan x = \sum_{k=0}^{\infty} \frac{(-1)^k x^{2k+1}}{2k + 1}, \qquad -1 \le x \le 1. \tag{39.9}$$

This formula for arctan x is known as *Gregory's Series,* after the Scotsman James Gregory, the original discoverer of Simpson's Rule (see Chap. 29). Gregory was *not* the original discoverer of his series, however, since essentially the same idea was known in India a century earlier. The special case of (39.9) for which $x = 1$ is the following interesting formula for π:

$$\frac{\pi}{4} = 1 - \frac{1}{3} + \frac{1}{5} - \frac{1}{7} + \frac{1}{9} - \cdots. \tag{39.10}$$

This formula is usually called *Leibniz' Series* because Leibniz discovered it independently (i.e., in ignorance of Gregory's work) a few years after Gregory's Series had been published (see Chap. 12 for further details). The problem with trying to use Leibniz' Series to compute π is that the series converges *very* slowly. As we observed in Chapter 12, the first real advance in the problem of computing π, after Archimedes' method of perimeters, was Gregory's Series; but to use it effectively, one must compute arctangents of numbers x that are as close to zero as possible. Here is a sample of how π can be related to arctangents of small numbers with the help of a trigonometric identity: Let $\alpha = \arctan \frac{1}{2}$ and $\beta = \arctan \frac{1}{3}$; then

$$\tan (\alpha + \beta) = \frac{\tan \alpha + \tan \beta}{1 - (\tan \alpha)(\tan \beta)}$$

$$= \frac{1/2 + 1/3}{1 - 1/6}$$

$$= 1$$

$$= \tan \frac{\pi}{4}.$$

Hence,
$$\frac{\pi}{4} = \alpha + \beta = \arctan \frac{1}{2} + \arctan \frac{1}{3}. \tag{39.11}$$

This formula was first published in 1776 by Charles Hutton, and was used by W. Lehmann in 1853 to compute π to 261 decimal places. A more extensive computation of π was carried out at about the same time by William Rutherford, using Machin's formula:

$$\frac{\pi}{4} = 4 \arctan \frac{1}{5} - \arctan \frac{1}{239}. \tag{39.12}$$

Machin's formula is older than Hutton's, having been published in 1706. It may be proved in a manner similar to the preceding proof for (39.11), but requiring several more steps. [See Hatcher for an elementary proof of this and several related arctangent formulas, and

Wrench (1960) for a detailed chronology of the use of such formulas in computations of π. Also see Chap. 12.]

39.3 A BETTER ALGORITHM [0]

Since our arctangent algorithm works only for $|x| \leq 1$, and rather badly for x near ± 1, it obviously wouldn't do for a built-in arctangent function. As in the case of the sine function (see Sect. 37.3), the way to remedy this difficulty is to compute arctan x in terms of the arctangent of some other number closer to zero. The starting point is a half-angle formula:

$$\tan \frac{1}{2} y = \frac{\sin y}{1 + \cos y} = \frac{\tan y}{\sec y + 1}$$

$$= \frac{\tan y}{1 + \sqrt{1 + \tan^2 y}} .$$

If we set $y = \arctan x$, this becomes

$$\tan \left(\frac{1}{2} \arctan x \right) = \frac{x}{1 + \sqrt{1 + x^2}} ,$$

or

$$\frac{1}{2} \arctan x = \arctan \frac{x}{1 + \sqrt{1 + x^2}} . \tag{39.13}$$

Thus, instead of computing arctan x directly, we may compute arctan $[x/(1 + \sqrt{1 + x^2})]$ and double it. That's a good start, because of the following fact (see Exercise 2 that follows):

$$\frac{|x|}{1 + \sqrt{1 + x^2}} < 1, \qquad \text{for all } x. \tag{39.14}$$

That really isn't good enough, though, because for large values of x we would have to use our arctangent algorithm on numbers near 1, thus risking very slow convergence. However, if we can determine *half* the desired angle (namely, arctan x), we can just as well determine *one-fourth* of it. If $y = x/(1 + \sqrt{1 + x^2})$, then

$$\frac{1}{4} \arctan x = \frac{1}{2} \arctan y = \arctan \frac{y}{1 + \sqrt{1 + y^2}} , \tag{39.15}$$

by double use of (39.13).

Exercise

2. Let $f(x) = x/(1 + \sqrt{1 + x^2})$, for all x. Show the following:
   a. f is increasing.
   b. $\lim_{x \to \pm\infty} f(x) = \pm 1$.

c. $|f(x)| < 1$ for all x.

d. If $y = f(x)$, then $|f(y)| < 1/(1 + \sqrt{2})$.

HINT: Use (a) and (c) to observe that it is necessary to check only $f(1)$ and $f(-1)$.

Now we have the ingredients of a really useful arctangent algorithm: For any x,

$$\arctan x = 4 \arctan z, \tag{39.16}$$

where

$$z = \frac{y}{1 + \sqrt{1 + y^2}} \quad \text{and} \quad y = \frac{x}{1 + \sqrt{1 + x^2}}. \tag{39.17}$$

We can compute $\arctan z$ by using approximating polynomials $A_1(z) = z$, $A_3(z)$, ..., knowing that $|z| < 1/(1 + \sqrt{2})$, and hence the error bound given by (39.6) will diminish rapidly. During your next session with the computer, you can check that $1/13(1 + \sqrt{2})^{13} < 0.0000009$ [this is the right-hand side of (39.6) for $n = 5$ and $x = 1/(1 + \sqrt{2})$]; so $A_{11}(z)$ will produce almost six-place accuracy in $\arctan z$ for $|z| < 1/(1 + \sqrt{2})$, and at least five-place accuracy for $\arctan x$ when computed as $4A_{11}(z)$. (Why?)

Exercise

3. Make a flowchart for and program the algorithm as previously described for computing $\arctan x$ for any x, using (39.16) and (39.17).

NOTE: This algorithm requires use of the built-in square root function; but this would pose no obstacle for the programmer who is designing a system like the one you use, since the system is at least as likely to contain a square root function as an arctangent function.

References

Hatcher; Lehmer; D. E. Smith; Wrench (1938, 1960).

C H A P T E R   4 0

COMPUTING π THE MODERN WAY (ca. A.D. 1700) [L]

Exercises

1. Check out your program from Exercise 1 of Chapter 39 by computing arctan x for several values of x with $|x| < 1$. Experiment with ϵ to get some feel for the trade-off between accuracy and degree as x gets farther away from zero. Also, compute arctan 1 $(=\pi/4)$ with $\epsilon = 0.01$. All answers may be checked by using the built-in arctangent function.

2. [O] If you have written a program for Exercise 3 of Chapter 39, use it instead. In this case, your testing should concentrate on getting reasonable agreement with the built-in arctangent over a wide range of x values. Also, check the value of $1/13(1 + \sqrt{2})^{13}$.

3. Compute $\pi/4$ by Hutton's formula (39.11):

$$\frac{\pi}{4} = \arctan \frac{1}{2} + \arctan \frac{1}{3}.$$

NOTE: The addition step may be done by hand, calculator, or computer.

4. Compute $\pi/4$ by Machin's formula (39.12):

$$\frac{\pi}{4} = 4 \arctan \frac{1}{5} - \arctan \frac{1}{239}.$$

NOTE: You will need a decimal approximation to $1/239$ for input purposes. You may compute this by hand, calculator, or computer.

5. Convert your arctangent program to a program to tabulate the arctangent function from $-\frac{1}{2}$ to $\frac{1}{2}$ in steps of 0.1, by deleting the input statement, setting the value of ϵ, modifying the output statement, and enclosing the rest of the program in a loop.

6. [O] If you have a program from Exercise 3 of Chapter 39, modify it as in Exercise 5 herein, to tabulate arctan x from -2 to 2 in steps of 0.2.

7. [O] If there is a numerical integration program available to you, use it to compute

$$\frac{\pi}{4} = \arctan 1 = \int_0^1 \frac{dt}{1 + t^2}.$$

C H A P T E R 4 1

HISTORICAL NOTE ON NATURAL LOGARITHMS [R]

The first problem we considered in this text was a compound-interest problem (Chap. 1). This problem has been of interest (no pun intended) for as long as money has been in use as a medium of exchange. One of the interesting finds that has turned up on a clay tablet from ancient Mesopotamia is an approximate solution to the following problem: How long will it take to double a sum of money invested at 20% interest, compounded annually? As you probably discovered in working through Chapters 1 and 2 (or may know from prior experience), if a sum P is invested for n years at compound interest rate r (expressed as a decimal fraction), the accumulated principal will be $P(1 + r)^n$. The Babylonian mathematicians knew this, and they wanted a solution of the equation

$$2P = P(1.20)^n,$$

or

$$2 = (1.2)^n, \tag{41.1}$$

which represents the (theoretical) time-to-doubling. Their solution was obtained by tabulating integral powers of 1.2, and performing a linear interpolation between $n = 3$ and $n = 4$ (see Fig. 41.1), which yielded a solution of 3.79 years (a rounded decimal equivalent of their sexagesimal fractions). Since you know about natural logarithms, you know that the exact solution of (41.1) can be expressed as $n = (\ln 2)/(\ln 1.2)$. A quick check of your handy log tables and a little arithmetic reveals a decimal approximation of slightly over 3.80, so the Babylonians didn't do so badly with their crude method. Of course, the solution to (41.1) can also be expressed as $\log_{1.2} 2$, essentially by definition. So the Babylonians were looking for logarithms, but didn't know it.

The impetus that gave rise to the invention of logarithms was the desire to convert computations of products into simpler computations of sums. However, this kind of computational trick was first accomplished with trigonometric functions rather than with logarithmic ones. We have already noted that the formulas for sines and cosines of sums and differences of angles were known to Ptolemy of Alexandria (see Chap. 37). From the Greek astronomer in Egypt (and elsewhere in the Greek world), trigonometry passed into the hands of the Hindu mathematicians in India, from whom it was subsequently learned by the Arabs, who brought it back to Egypt (and elsewhere in the Arab world). By 1000 A.D. it was known in Egypt that

$$\cos A \cos B = \frac{1}{2} [\cos (A + B) + \cos (A - B)], \tag{41.2}$$

Figure 41.1

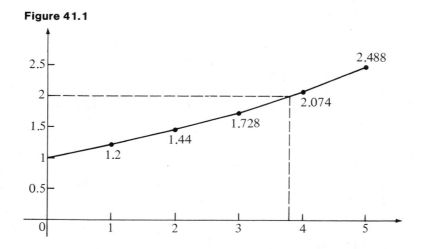

a formula that expresses the product of two numbers as the sum of two other numbers. Although (41.2) seems only a small step from the formulas known to Ptolemy, it was not until the sixteenth century that it and similar formulas were widely used for computing products. Of course, we would be a little hampered in using (41.2) by our hangup about cosines being less than 1 (in absolute value); but the mathematicians and astronomers of the sixteenth century were under no such illusion. Nevertheless, if we had to multiply 9482 and 287, it would be no harder to multiply 0.9482 and 0.287 instead (both of these numbers being cosines of something), and then discard the decimal point. Since tables of sines and cosines to a dozen or more places were common in the sixteenth century, the use of such formulas for products became popular.

Late in that same century, John Napier (1550–1617), a Scottish laird who dabbled in mathematics for fun, not for profit, spent about 20 years perfecting the idea of the Babylonians: tabulating powers. Napier's objective was to use the properties of exponents to convert products into sums, and also to get his powers close enough together so that linear interpolation between them would be accurate. Realizing that the powers would quickly spread out unless they were powers of a number close to 1, he chose to tabulate powers of $1 - 10^{-7}$ (multiplied by a scale factor of 10^7 to shift the decimal point). Thus his idea of the "logarithm" of a number N was a number L such that $10^7(1 - 10^{-7})^L = N$. Since, as we see in a later chapter, $(1 - 10^{-7})^{10^7}$ is very close to $1/e$ (where e is the base for natural logarithms), Napier's system was essentially equivalent to logarithms with base $1/e$. However, he had no concept of either *base* or the number e. He did, though, coin the word we know this system by, from the Greek words *logos* (ratio) and *arithmos* (number). The results of his 20 years of labor were published in Latin in 1614 under the title, "A Description of the Marvelous Rule of Logarithms."

Napier was not alone in developing the idea of logarithm. Similar ideas were developed independently and at about the same time by a Swiss clockmaker, Jobst Bürgi (1552–1632). In fact, Bürgi may have had the idea first, but his results were not published until 1620, so Napier was accorded priority. However, Bürgi's system was actually closer to our "Napierian," or natural, logarithms, since he used powers of $(1 + 10^{-4})$, the 10^4-th power of which is close to e.

In the next chapter we derive an infinite-series expression (i.e., a sequence of polynomial approximations) for log $(1 + x)$ in powers of x. (The shift of 1 is necessary because one cannot reasonably expect to calculate log x near $x = 0$.) This expression is known as *Mercator's Series,* after the Danish mathematician Nicolaus Mercator (1620–1687), whose real name was Kaufmann [not to be confused with the Flemish geographer Gerard Mercator (1512–1594), whose real name was Kremer]. The same idea may have been known earlier to Sir Isaac Newton and to the Dutch mathematician Johann Hudde (1629–1704). The derivation we give, using integration of the terms of a geometric progression (similar to the derivation of the arctangent formula in Chap. 39), was published by Leibniz in 1677.

The concept of base for logarithms and the use of the number e came much later. In 1748 Léonhard Euler published an infinite-series expression for exponential functions with arbitrary base, a special case of which we study later (base e). He also introduced the symbol e for the natural base and calculated its value to 18 places, using a limit formula that will be the subject of our subsequent chapter on compound interest. Euler tried to show that e is irrational, but without complete success. More than a century later, the Frenchman Charles Hermite (1822–1901) showed, by a very complicated argument, that e is not even algebraic (i.e., not a root of any algebraic equation). (See Pennisi for a very simple proof of the irrationality of e, and Schenkman for a proof of the much more difficult fact that e is not algebraic.)

Like π, extended decimal approximations to e have attracted the interest of many skilled computers, including electronic ones. An approximation to 60,000 decimal places was computed by D. Wheeler in 1952, and this was extended to 100,000 places by Daniel Shanks in 1961 (see Stoneham).

References

Boyer (1968); Cairns; Cajori (1913); Coolidge (1950); Eves (1960); Gardner (1960b, 1973a); Higgins; Pennisi; Read (1960); Schenkman; Stoneham; von Baravelle (1945).

C H A P T E R   4 2

THE NATURAL LOGARITHM FUNCTION

In this chapter we derive a sequence of polynomial approximations to $\ln(1 + x)$ and show how this sequence can be used to calculate the natural log of any positive number. Our starting point is the usual definition:

$$\ln(1 + x) = \int_1^{1+x} \frac{dt}{t} = \int_0^x \frac{du}{1 + u}, \tag{42.1}$$

where $u = t - 1$. Recall the formula[1]

$$1 + r + r^2 + \cdots + r^{n-1} = \frac{1 - r^n}{1 - r},$$

for any number r and any positive integer n. If we set $r = -u$, this becomes

$$1 - u + u^2 - \cdots + (-1)^{n-1}u^{n-1} = \frac{1}{1 + u} - \frac{(-1)^n u^n}{1 + u}.$$

Substituting the resulting expression for $1/(1 + u)$ in (42.1), we get

$$\ln(1 + x) = \int_0^x \left[1 - u + u^2 - \cdots + (-1)^{n-1}u^{n-1} + \frac{(-1)^n u^n}{1 + u} \right] du$$

$$= x - \frac{x^2}{2} + \frac{x^3}{3} - \cdots + \frac{(-1)^{n-1}x^n}{n} + \int_0^x \frac{(-1)^n u^n}{1 + u}\, du.$$

Let's abbreviate the polynomial part of the previous expression:

$$L_n(x) = x - \frac{x^2}{2} + \frac{x^3}{3} - \cdots + (-1)^{n-1}\frac{x^n}{n}. \tag{42.2}$$

Then we have, for each positive integer n,

$$\ln(1 + x) = L_n(x) + \int_0^x \frac{(-1)^n u^n}{1 + u}\, du. \tag{42.3}$$

Now we need to estimate the difference between $\ln(1 + x)$ and $L_n(x)$, represented by the absolute value of the integral in (42.3). If we confine our attention to positive values of x,

---

[1] This may be obtained by dividing $(1 - r)$ into 1 by "long division," with remainder [cf. Sect. 39.2]; or see the first half of the proof of Theorem 11.2.

then for all u in the interval $[0, x]$ we have $1 + u \geq 1$, so

$$\int_0^x \frac{u^n}{1 + u}\, du \leq \int_0^x u^n\, du = \frac{x^{n+1}}{n + 1}.$$ (42.4)

Hence,

$$|\ln(1 + x) - L_n(x)| = \int_0^x \frac{u^n}{1 + u}\, du \leq \frac{x^{n+1}}{n + 1},$$ (42.5)

for $x \geq 0$ and any positive integer n. In particular, if $x \leq 1$, then $x^{n+1}/(n+1) \leq 1/(n+1)$, which converges to 0 as n becomes large, so we have

$$\ln(1 + x) = \lim_{n \to \infty} L_n(x) \qquad \text{for } 0 \leq x \leq 1.$$ (42.6)

[Written in infinite-series notation, (42.6) is *Mercator's Series* (see Chap. 41):

$$\ln(1 + x) = \sum_{n=1}^{\infty} \frac{(-1)^{n-1} x^n}{n}, \qquad 0 \leq x \leq 1.]$$

However, that's not really enough for a practical algorithm for computing logarithms because the interval on which convergence is assured is much too small, and the error estimate (42.5) suggests rather slow convergence when x is close to 1. To rectify this situation, we use the properties of the log function and some outright tricks.

First, let's replace x by $-x$ in (42.3), which did not depend on x being positive:

$$\ln(1 - x) = L_n(-x) + \int_0^{-x} \frac{(-1)^n u^n}{1 + u}\, du.$$ (42.7)

Replacement of x by $-x$ in (42.2) yields

$$L_n(-x) = -x - \frac{x^2}{2} - \frac{x^3}{3} - \cdots - \frac{x^n}{n}.$$ (42.8)

Subtracting (42.7) from (42.3), we have

$$\ln\left(\frac{1 + x}{1 - x}\right) = \ln(1 + x) - \ln(1 - x)$$

$$= L_n(x) - L_n(-x) + \int_0^x \frac{(-1)^n u^n}{1 + u}\, du - \int_0^{-x} \frac{(-1)^n u^n}{1 + u}\, du.$$ (42.9)

This suggests the use of a new sequence of polynomials for approximating $\ln[(1 + x)/(1 - x)]$, namely,

$$L_n(x) - L_n(-x) = 2x + \frac{2x^3}{3} + \frac{2x^5}{5} + \cdots + \frac{2x^{2k-1}}{2k - 1},$$

where n is either $2k$ or $(2k - 1)$. (Note that all even-power terms cancel in the subtraction step.) Let's abbreviate these new polynomials:

$$P_{2k-1}(x) = 2\left[x + \frac{x^3}{3} + \frac{x^5}{5} + \cdots + \frac{x^{2k-1}}{2k - 1}\right].$$ (42.10)

Then (42.9) may be rewritten as

$$\ln\left(\frac{1+x}{1-x}\right) = P_{2k-1}(x) + \int_0^x \frac{(-1)^{2k-1}u^{2k-1}}{1+u}\,du + \int_0^{-x} \frac{(-1)^{2k}u^{2k-1}}{1+u}\,du; \quad (42.11)$$

hence

$$\left|\ln\left(\frac{1+x}{1-x}\right) - P_{2k-1}(x)\right| \le \int_0^x \frac{u^{2k-1}}{1+u}\,du + \left|\int_0^{-x} \frac{u^{2k-1}}{1+u}\,du\right|, \quad (42.12)$$

for $x \ge 0$. The first integral on the right has already been estimated [set $n = 2k - 1$ in (42.4)]. In particular, if we restrict attention to $0 \le x \le \frac{1}{2}$, we have, from (42.4),

$$\int_0^x \frac{u^{2k-1}}{1+u}\,du \le \frac{1}{(2k)2^{2k}} = \frac{1}{k2^{2k+1}}. \quad (42.13)$$

To estimate the second integral, let's make the substitution $u = -z$. Then

$$\left|\int_0^{-x} \frac{u^{2k-1}}{1+u}\,du\right| = \left|\int_0^x \frac{z^{2k-1}}{1-z}\,dz\right|.$$

(Factors of -1 have been ignored because they do not change the absolute value.) Now, for values of x in the interval $[0, \frac{1}{2}]$ and $0 \le z \le x$, we have $1 - z \ge \frac{1}{2}$, so

$$\left|\int_0^x \frac{z^{2k-1}}{1-z}\,dz\right| \le 2 \int_0^x z^{2k-1}\,dz = \frac{x^{2k}}{k} \le \frac{1}{k2^{2k}}. \quad (42.14)$$

Combining (42.12) with the estimates (42.13) and (42.14), we get

$$\left|\ln\left(\frac{1+x}{1-x}\right) - P_{2k-1}(x)\right| \le \frac{3}{k2^{2k+1}}. \quad (42.15)$$

You may have wondered by now just what's going on here—our problem with the L's was that they were guaranteed to work only for $0 \le x \le 1$, but we have just devoted considerable effort to getting an error estimate for the P's that applies only for $0 \le x \le \frac{1}{2}$. The important difference that makes the P's more useful lies in the set of numbers for which we can compute logarithms by using the P's, namely, all numbers of the form

$$y = \frac{1+x}{1-x}, \qquad 0 \le x \le \frac{1}{2}. \quad (42.16)$$

Exercises

1. Sketch the graph of the function defined by (42.16). In particular, note that the function is increasing and that y ranges over the interval $[1, 3]$. [It follows that the P's are useful for computing logarithms over an interval at least twice as large as the one where the L's would work, and with no "slow convergence" problems because the right-hand member of (42.15) decreases rapidly to 0 as k becomes large.]

2. Solve (42.16) for x to obtain the inverse function

$$x = \frac{y-1}{y+1}, \qquad 1 \le y \le 3. \quad (42.17)$$

3. Make a flowchart for the following algorithm for computing ln y on the interval $[1, 3]$ to within ϵ. Replace y by x computed from (42.17). Then compute $P_1(x) = x$, $P_3(x)$, $P_5(x), \ldots$, using a recursive version of (42.10):

$$P_{2k+1}(x) = P_{2k-1}(x) + \frac{2x^{2k+1}}{2k + 1} .$$ (42.18)

Stop when the error estimate (42.15) assures the desired accuracy. Your output should include the last $P(x)$ (approximately ln y) and the degree of the polynomial used for the answer. Write a program to implement the flowchart.

4. **a.** What property of ln ensures that logarithms of numbers in $(0, 1)$ can be computed from logarithms of corresponding numbers greater than 1?
   b. If $x > 1$ and n is a positive integer such that $e^n \le x \le e^{n+1}$, show that

$$\ln x = n + \ln y,$$ (42.19)

   where y is a number in $[1, 3]$.

HINT: Divide through by e^n.

5. [O] Modify your program from Exercise 3, using the ideas in Exercise 4, to create a program that computes ln x for *any* $x > 0$. You need an accurate value for e in this program to locate the appropriate n for (42.19):

$$e = 2.718281828459045$$

should do. (Later we will see several ways to compute e.) If you want to design this program with a fixed error tolerance, you can observe that

$$\frac{3}{8 \cdot 2^{17}} = \frac{3}{2^{20}} < 0.000003,$$

since $2^{10} = 1024 > 10^3$, and therefore (42.15) ensures that P_{15} provides five-place accuracy for all x.

References

Ballantine; Kuller.

C H A P T E R 4 3

LET'S COMPUTE LOGARITHMS [L]

Exercises

1. Check out your program from Exercise 3 of Chapter 42 by computing several values of ln x for $1 \leq x \leq 3$. Experiment with ϵ and note the degrees of polynomials required to achieve various levels of accuracy at different points in the interval. All answers may be checked by using the built-in natural log function.

2. [O] If you have written a program for Exercise 5 in Chapter 42, use it instead. In this case, your testing should concentrate on getting reasonable agreement with the built-in log function over a wide range of positive numbers.

3. Solve the time-to-doubling problem of the ancient Babylonians (see Chap. 41), i.e., solve the equation $2 = (1.2)^n$ for n.

4. Solve the time-to-doubling problem of Chapter 1: 8.8% interest, compounded quarterly.

5. Convert your program to tabulate the natural log function from 1 to 3, in steps of 0.1: Delete the input statement, set the value of ϵ, modify the output statement, and enclose the rest of the program in an appropriate loop.

6. If you have a numerical integration program available, use it to compute ln 2 and ln 3 from the definition: ln $x = \int_1^x (dt/t)$.

7. [O] If you have a program from Chapter 21 available, use it to minimize each of the following functions:
   a. $f(x) = x \ln x$ on [0.1, 1].
   b. $f(x) = \ln [1 - \arctan (x/2)]$ on [0, 1].

   (Alternatively, take derivatives and use a root-finding program from Chap. 15 or 18.)

C H A P T E R 4 4

POWER SERIES: A LOOK AHEAD AND A LOOK BACKWARD [R]

The seven preceding chapters have all centered around the theme of polynomial approximations to nonpolynomial functions. We have also noted that these approximations can be expressed as representations of the given functions by infinite series of powers (or *power series*), which are really just limits of sequences of polynomials. Our polynomial approximations have all been designed to "fit" their respective functions quite well near $x = 0$; but the same ideas could be applied to obtain good fits near any value $x = a$. The difference is that the resulting series would involve powers of $(x - a)$ instead of powers of x.

All the results obtained concerning these polynomial approximations are special cases of the following theorem, known as Taylor's Theorem,[1] which you will study later in your calculus course:

Let f be a function that is infinitely differentiable on an interval I containing the numbers a and x. Then, for each positive integer n, the value of f at x is given by

$$f(x) = f(a) + f'(a)(x - a) + \frac{f''(a)}{2!}(x - a)^2 + \frac{f'''(a)}{3!}(x - a)^3$$

$$+ \cdots + \frac{f^{(n)}(a)}{n!}(x - a)^n + R_n(x, a),$$

where

$$R_n(x, a) = \frac{1}{n!}\int_a^x (x - t)^n f^{(n+1)}(t)\, dt$$

$$= \frac{(x - a)^{n+1} f^{(n+1)}(c)}{(n + 1)!}$$

for some number c between x and a.

In any situation in which it can be shown that

$$\lim_{n \to \infty} R_n(x, a) = 0,$$

using either of the given forms for $R_n(x, a)$, it follows that

$$f(x) = \sum_{n=0}^{\infty} \frac{f^{(n)}(a)}{n!}(x - a)^n. \tag{44.1}$$

---

[1]See also (optional) Chapter 20, where the special case of this theorem for $n = 2$ was stated and used.

The right-hand side of (44.1) is called a *Taylor series* representation for $f(x)$. It is a power series in powers of $(x - a)$ because the coefficients $f^{(n)}(a)/n!$ are constants—i.e., they don't depend on x. The partial sums of this series—i.e., the polynomials that appear in Taylor's Theorem—are called *Taylor polynomials* for f.

An important special case is that for which $a = 0$, in which the Taylor series and polynomials are called, respectively, *Maclaurin series* and *Maclaurin polynomials*. For example, if $f(x) = \sin x$, then its successive derivatives are $\cos x$, $-\sin x$, $-\cos x$, $\sin x$, and then this pattern repeats. The values of these functions at $x = 0$ (starting with f) are 0, 1, 0, -1, 0, and so on; so Taylor's Theorem with $a = 0$ leads to the Maclaurin polynomials $S_{2n+1}(x)$ of (37.13) and the Maclaurin series (37.25).

Following a pattern we have observed many times before, Taylor's Theorem was not first proved by the man it is named for, nor are Maclaurin series due to Maclaurin. As we noted in Chapter 39, the idea of an infinite-series representation of a function was known in India in the sixteenth century. The Taylor series representation was discovered by James Gregory and published in 1668 (long before Taylor was born). Gregory knew the Maclaurin series for $\tan x$, $\sec x$, $\arctan x$, and $\operatorname{arcsec} x$, and (as we have noted) one of these is actually named for him. It was at about the same time (30 years before Maclaurin was born) that Nicolaus Mercator discovered the Maclaurin series for $\ln(1 + x)$ (see Chaps. 41 and 42).

Brook Taylor (1685–1731) was an English mathematician who in 1715 published a book entitled *Methodus incrementorum directa et inversa,* which contained the representation we know today as Taylor series, but he was apparently unaware of the earlier work of Gregory. It was not until 1742 that the Scotsman Colin Maclaurin (1698–1746) published his *Treatise of Fluxions,* in which he quoted Taylor's work and made no claim to any new result on series representations. Nevertheless, his role as author was confused by his contemporaries with that of discoverer, and we have, in a sense, inherited that confusion. Maclaurin was, however, a brilliant mathematician whose best work was in geometry. He was a professor of mathematics at Aberdeen at the age of 19, and he was also the original discoverer of the method for solving simultaneous linear equations that we know today as "Cramer's Rule."

References

Boyer (1968); Klein (pp. 233–234); Sanford (1934a, 1934b); Wollan.

C H A P T E R   4 5

A CONSUMER'S GUIDE TO COMPOUND INTEREST

45.1 FREQUENT COMPOUNDING

A popular advertising ploy of banks is to announce that acquisition of appropriate computer technology now makes it possible for them to compound your interest *daily,* rather than quarterly or at some longer interval. Along with such an announcement may come the news that 9.8% interest, say, compounded daily, produces an "effective yield" of 10.3%, whatever that means.

In this chapter we study the significance of more frequent compounding of interest, and also discover a connection with the way Napier and Bürgi computed "logarithms" and Euler computed e (see Chap. 41).

Let's write $f(t)$ for a bank balance at the end of t years (not necessarily an integral number of years), starting with an initial deposit $f(0)$, at an interest rate of $100x$ percent (i.e., x is the decimal fraction corresponding to the interest rate), compounded k times per year. Our problem is to determine $f(t)$ as an explicit function of t, involving, of course, the given constants k, x, and $f(0)$.

The interval for each compounding is $1/k$, and the interest rate for that period, expressed as a decimal fraction, is x/k. Thus we have

$$f\left(\frac{1}{k}\right) = f(0) + \frac{x}{k}f(0) = \left(1 + \frac{x}{k}\right)f(0),$$

$$f\left(\frac{2}{k}\right) = f\left(\frac{1}{k}\right) + \frac{x}{k}f\left(\frac{1}{k}\right) = \left(1 + \frac{x}{k}\right)^2 f(0),$$

$$f\left(\frac{3}{k}\right) = f\left(\frac{2}{k}\right) + \frac{x}{k}f\left(\frac{2}{k}\right) = \left(1 + \frac{x}{k}\right)^3 f(0),$$

and so on. Evidently, each new compounding increases the balance by a factor of $(1 + x/k)$. Hence for each time t of the form $t = m/k$ (or $kt = m$, a positive integer),

$$f(t) = \left(1 + \frac{x}{k}\right)^{kt} f(0). \tag{45.1}$$

In particular, if interest is compounded quarterly, the bank balance is given by

$$f(t) = \left(1 + \frac{x}{4}\right)^{4t} f(0), \tag{45.2}$$

whereas compounding daily gives

$$f(t) = \left(1 + \frac{x}{365}\right)^{365t} f(0), \tag{45.3}$$

and compounding every *second* gives

$$f(t) = \left(1 + \frac{x}{31,536,000}\right)^{31,536,000t} f(0). \tag{45.4}$$

All these formulas are of the form

$$f(t) = a_k^t f(0), \tag{45.5}$$

where

$$a_k = \left(1 + \frac{x}{k}\right)^k. \tag{45.6}$$

Thus the study of more frequent compounding essentially comes down to the study of the sequence $\{a_k\}$ as k becomes large. (Keep in mind that x is a constant.) Now it seems evident that more frequent compounding is at least favorable to the depositor; i.e., the sequence defined by (45.6) should be *increasing,* although that is by no means evident from formula (45.6). But note also that as the exponent k gets very large, x/k gets very small, so the number being raised to a large power is very close to 1. It is therefore not at all clear what the limit is as k goes to infinity (i.e., the ultimate compounding of interest).

If you recall the results of Section 10.2, you might be prepared to make some guesses about $\lim_{k \to \infty} a_k$, at least for certain values of x. In particular, Exercise 2 of that section involved computing terms of the sequence for $x = 1$ (meaning 100% interest), and Exercise 4 dealt with $x = -1$ (which of course would not correspond to an acceptable interest rate, at least from the depositor's point of view).

Exercise

1. For use in the next laboratory session, prepare a very simple program that will accept x and k as inputs and print out $(1 + x/k)^k$. Note from (45.5) and (45.6) that this expression is just $f(1)/f(0)$, the factor by which your money grows in one year when compounded k times per year.

45.2 CONTINUOUS COMPOUNDING

Better evidence about the nature of the limit we seek is obtained by examining the notion of continuous compounding of interest, which has also been used by some banks. This means that $f(t)$ is a continuously changing function whose *rate* of change is given by

$$f'(t) = xf(t). \tag{45.7}$$

Remembering that x is constant, you should recognize (45.7) as the differential equation for exponential growth. Its solution is

$$f(t) = e^{xt}f(0). \tag{45.8}$$

Comparison of the results for frequent compounding, (45.1) through (45.5), and that for continuous compounding, (45.8), suggests the possibility that

$$\lim_{k \to \infty} \left(1 + \frac{x}{k}\right)^k = e^x. \tag{45.9}$$

In addition to our announced computational goal of determining how much difference frequent compounding makes, we now consider the mathematical goal of showing that (45.9) is correct for every x.

The latter task may be approached indirectly, using some facts from your calculus course. As you know, the derivative of ln x is $1/x$, so the slope of the natural log function at $x = 1$ is 1. Expressed as a limit (i.e., using the definition of derivative), this says that

$$1 = \lim_{\Delta x \to 0} \frac{\ln (1 + \Delta x) - \ln 1}{\Delta x} = \lim_{\Delta x \to 0} \frac{\ln (1 + \Delta x)}{\Delta x}.$$

Multiplying through by x, we have

$$x = \lim_{\Delta x \to 0} \frac{x \ln (1 + \Delta x)}{\Delta x}.$$

Now set $k = x/\Delta x$, and consider only those values of Δx such that k is a positive integer. As Δx goes to 0, k goes to infinity, and we have

$$x = \lim_{k \to \infty} k \ln \left(1 + \frac{x}{k}\right) = \lim_{k \to \infty} \ln \left(1 + \frac{x}{k}\right)^k.$$

Finally, we take exp of both sides,[1] and use the fact that the exponential function is continuous:[2]

$$e^x = \exp\left[\lim_{k \to \infty} \ln \left(1 + \frac{x}{k}\right)^k\right]$$

$$= \lim_{k \to \infty} \left[\exp \ln \left(1 + \frac{x}{k}\right)^k\right]$$

$$= \lim_{k \to \infty} \left(1 + \frac{x}{k}\right)^k,$$

because $e^{\ln y} = y$ for any positive number y. This completes the proof of (45.9).

The limit formula (45.9) gives a way of actually computing values of e^x. However, it is not an efficient way because the sequence actually converges rather slowly. Nevertheless, as we

---

[1] We are using exp as a name for the exponential function with base e: exp $(x) = e^x$.

[2] [O] The formal justification for interchanging "exp" and "lim" can probably be found in your calculus book. The argument is very similar to the proof of Theorem 11.3 in this book.

noted in Chapter 41, Euler computed e to 18 places by using the case of this formula for $x = 1$:

$$e = \lim_{k \to \infty} \left(1 + \frac{1}{k} \right)^k. \tag{45.10}$$

Furthermore, formula (45.9) also verifies the assertions made in Chapter 41 in connection with "Napierian" logarithms:

$$\left(1 - \frac{1}{10^7} \right)^{10^7} \text{ is close to } e^{-1}$$

and

$$\left(1 + \frac{1}{10^4} \right)^{10^4} \text{ is close to } e.$$

Exercise

2. Write a program to compute terms of the sequence $\{a_k\}$ defined by (45.6) for $k = 1, 2, 4, 8,$ ... (i.e., double k each time through the loop). Stop when the difference of successive terms is less than ϵ. Consider x and ϵ as input variables, and print all computed terms of the sequence.

NOTE: This program is very similar to the one for Section 10.2.

45.3 POLYNOMIAL APPROXIMATION TO THE EXPONENTIAL FUNCTION [O]

The program called for in Exercise 2 is one way to compute e^x, according to formula (45.9), but as previously noted, it is not a very good way. It is possible to obtain polynomial approximations to e^x by using a method that is very similar to that used in Chapter 37 for the sine and cosine functions, which is a much better computational method. Later (Chap. 56), when we can draw on more information from your calculus course, we consider a still better method.

For technical reasons that will be more apparent later, it is convenient to look for approximations to e^{-x} rather than to e^x. This presents no special problem, however, because $e^x = 1/e^{-x}$. Our starting point is to consider the tangent line to $y = e^{-x}$ at the point $(0, 1)$. You may check that this line has the equation $y = 1 - x$ (do it!). Now the graph of $y = e^{-x}$ is concave upward everywhere. (Why?) So any tangent line must lie entirely *below* the graph, except at the point of tangency. Hence

$$e^{-x} \geq 1 - x, \qquad \text{for all } x. \tag{45.11}$$

Now let's change the variable in (45.11) to t and integrate each member from 0 to x:

$$\int_0^x e^{-t} \, dt = -e^{-t} \Big|_0^x = 1 - e^{-x};$$

$$\int_0^x (1 - t) \, dt = t - \frac{t^2}{2} \Big|_0^x = x - \frac{x^2}{2}.$$

If $x \geq 0$, the inequality (45.11) is preserved when both sides are integrated, and we have

$$1 - e^{-x} \geq x - \frac{x^2}{2},$$

or

$$e^{-x} \leq 1 - x + \frac{x^2}{2}, \qquad x \geq 0. \tag{45.12}$$

Exercise

3. Repeat the process of replacing x by t, integrating from 0 to x, and solving for e^{-x} to show that

$$e^{-x} \geq 1 - x + \frac{x^2}{2} - \frac{x^3}{6}, \qquad x \geq 0, \tag{45.13}$$

and

$$e^{-x} \leq 1 - x + \frac{x^2}{2} - \frac{x^3}{6} + \frac{x^4}{24}, \qquad x \geq 0. \tag{45.14}$$

At this point we should introduce abbreviations for the polynomials that are appearing as upper and lower bounds for e^{-x}:

$$E_n(x) = 1 - x + \frac{x^2}{2!} - \frac{x^3}{3!} + \cdots + \frac{(-1)^n x^n}{n!}, \qquad n = 1, 2, 3, \cdots. \tag{45.15}$$

What we have discovered so far is that $E_1(x)$ and $E_3(x)$ are $\leq e^{-x}$ for $x \geq 0$, and $E_2(x)$ and $E_4(x)$ are $\geq e^{-x}$ for $x \geq 0$. Obviously this process continues indefinitely: *All* the $E_n(x)$ for n odd lie below e^{-x}, and for n even lie above e^{-x}. If we write the definition (45.15) recursively:

$$E_{n+1}(x) = E_n(x) + \frac{(-1)^{n+1} x^{n+1}}{(n+1)!}, \tag{45.16}$$

then we can express the difference of consecutive terms of the sequence (for each fixed, positive x) by

$$\left| E_{n+1}(x) - E_n(x) \right| = \frac{x^{n+1}}{(n+1)!}. \tag{45.17}$$

Since e^{-x} lies *between* $E_n(x)$ and $E_{n+1}(x)$ for any n, we have

$$\left| e^{-x} - E_n(x) \right| \leq \frac{x^{n+1}}{(n+1)!}, \qquad x \geq 0. \tag{45.18}$$

By applying Theorem 37.2, we also see from (45.18) that

$$e^{-x} = \lim_{n \to \infty} E_n(x) = \sum_{n=0}^{\infty} \frac{(-1)^n x^n}{n!}, \qquad x \geq 0. \qquad (45.19)$$

Exercises

4. Make a flowchart for, and program, an algorithm for computing e^y for any y, to within ϵ. If $y < 0$, say, $y = -x$, then calculate e^{-x} as previously described. Otherwise, calculate e^{-y} as before, and use $e^y = 1/e^{-y}$. Note that the error bound in (45.18) is the absolute value of the next term to be added, according to (45.16). Also note that the new term in (45.16) is $-x/(n + 1)$ times the last previous term added (cf. Exercise 37.1). Include in your output the degree of the approximating polynomial.

5. The approximating polynomials $E_n(x)$ converge to e^{-x} for *every* value of x; but as with the approximations to $\sin x$ and $\cos x$, the convergence is slow when x is large. Unfortunately, e^{-x} does not have the periodic nature of the sine and cosine that allowed replacement of a large x by a smaller one having the same sine or cosine. Can you think of a way to use the properties of the exponential function to compute e^{-x} when x is large, by using $E_n(z)$ for some z near 0?

6. The infinite-series expression (45.19) for e^{-x} bears a striking similarity to the expressions (37.25) and (37.26) for $\sin x$ and $\cos x$. This suggests that these three functions may be related somehow. Can you figure out from the series expressions what the relation among them might be?

WARNING: The answer to this question lies outside the world of real numbers in which you have operated throughout your calculus course.

C H A P T E R   4 6

EFFECTIVE INTEREST RATES AND e^x [L]

Exercises

1. Use your program from Exercise 1 of Chapter 45 to compare annual growth rates for 9.8% interest ($x = 0.098$) compounded every quarter ($k = 4$), every day ($k = 365$), and every second ($k = 31536000$). What is the meaning of the claim that compounding daily produces an "effective yield" of 10.3%?

2. Use your program from Exercise 2 of Chapter 45 to compute e^x for $x = 1, 2, 3, 0.5, 0.098,$ $-1, -2$ (and others if you wish). Compare your answers with values obtained from the built-in exponential function. An ϵ on the order of 10^{-4} will work in most cases, but for small values of x, you may be able to use a slightly smaller value. Also, compare $e^{0.098}$ with the answers obtained in Exercise 1.

3. Suppose that bank A offers $6\frac{1}{4}\%$ interest compounded daily, bank B offers $6\frac{3}{8}\%$ compounded quarterly, and bank C offers $6\frac{1}{2}\%$ compounded annually. Which bank pays more interest on a given deposit after an integral number of years? If you had a large sum to deposit, can you think of a reason for splitting your deposit among the banks?

4. [O] Check out your program from Exercise 4 of Chapter 45 by computing several values of e^x for both positive and negative x. Experiment with ϵ, and take note of the degrees of polynomials required to achieve various levels of accuracy for various sizes of x. All answers may be checked by using the built-in exponential function.

5. [O] Convert your program from Exercise 4, Chapter 45, to a program to tabulate the exponential function from -5 to 5 in steps of 0.5 with $\epsilon = 10^{-5}$.

C H A P T E R   4 7

IS MORE FREQUENT COMPOUNDING BETTER? [O]

The discussion of compound interest in Chapter 45 should have made it intuitively obvious that, for each positive number x, the sequence

$$a_k = \left(1 + \frac{x}{k} \right)^k \tag{47.1}$$

is increasing. Namely, think of x as an interest rate and k as the number of times per year the interest is compounded: More frequent compounding should yield more interest. We now proceed to see why this is so.

Our starting point is to look at the graph of $f(t) = \ln (1 + t)$ (see Fig. 47.1). Note that $f'(t) = 1/(1 + t)$ is a decreasing function of t. Consideration of the chords of the graph from 0 to b and from 0 to c (where $0 < b < c$) would make it appear that their slopes are in the relation

$$\frac{f(b)}{b} > \frac{f(c)}{c} . \tag{47.2}$$

This is not immediately obvious from the definition of f, but it is obvious that the slope $f(b)/b$ from 0 to b is larger than the slope from b to c—i.e.,

$$\frac{f(b)}{b} > \frac{f(c) - f(b)}{c - b} . \tag{47.3}$$

Figure 47.1

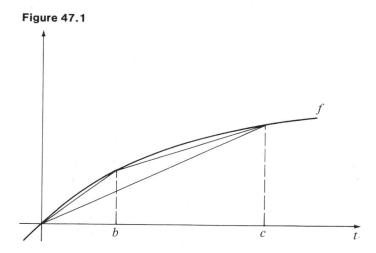

REASON: Each of these is a value of f', by the Mean Value Theorem, the first on $(0, b)$, the second on (b, c), and f' is a decreasing function.

Now multiply out the fractions in (47.3) and cancel like terms on both sides of the inequality. The result is

$$cf(b) > bf(c), \qquad (47.4)$$

which is equivalent to (47.2).

Now let x be any positive number, k any positive integer, and set $b = x/(k + 1)$, $c = x/k$ in (47.4):

$$\frac{x}{k} \ln\left(1 + \frac{x}{k + 1} \right) > \frac{x}{k + 1} \ln\left(1 + \frac{x}{k} \right). \qquad (47.5)$$

Multiply both sides of (47.5) by $k(k + 1)/x$ (which is positive), and take exp of both sides (thus preserving the inequality, as exp is an increasing function). The result is

$$\left(1 + \frac{x}{k + 1} \right)^{k+1} > \left(1 + \frac{x}{k} \right)^k, \qquad (47.6)$$

which shows that the sequence defined by (47.1) is indeed increasing.

Exercises

1. Let f be any function such that f'' is continuous and negative throughout an interval $[a, c]$. Let b be any number between a and c. Show that

   a. $\dfrac{f(b) - f(a)}{b - a} > \dfrac{f(c) - f(b)}{c - b}$;

   b. $(c - a)f(b) > (c - b)f(a) + (b - a)f(c)$.

NOTE: Our preceding computation is a special case of this result. See Bush for additional applications.

2. Consider the graph of $f(t) = 1/t$ on the interval $[k, k + x]$, where k is a positive integer and x is any positive number (see Fig. 47.2). Show

   a. $\dfrac{x}{k} > \displaystyle\int_k^{k+x} \dfrac{dt}{t} > \dfrac{x}{k + x}$;

   b. $\displaystyle\int_k^{k+x} \dfrac{dt}{t} = \ln\left(1 + \dfrac{x}{k} \right)$;

   c. $x > k \ln\left(1 + \dfrac{x}{k} \right) > \dfrac{kx}{k + x}$;

   d. $e^x > \left(1 + \dfrac{x}{k} \right)^k > e^{x/(1 + x/k)}$;

   e. $\displaystyle\lim_{k \to \infty} \left(1 + \dfrac{x}{k} \right)^k = e^x$.

NOTE: This is an alternative proof of (45.9), based on the note by Schaumberger.

Figure 47.2

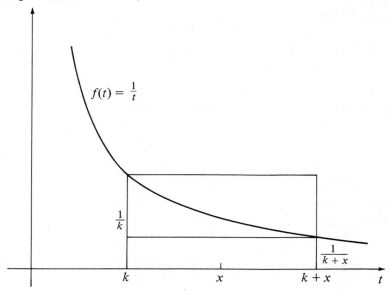

$f(t) = \frac{1}{t}$

3. Use Exercise 2(a) and (b) to show

   a. $(k + x) \ln \left(1 + \dfrac{x}{k} \right) > x;$

   b. $\left(1 + \dfrac{x}{k} \right)^{k+x} > e^x.$

   In particular, for $x = 1$, we have

   c. $\left(1 + \dfrac{1}{k} \right)^{k} < e < \left(1 + \dfrac{1}{k} \right)^{k+1}$

   Use this to show that

   d. $\left| e - \left(1 + \dfrac{1}{k} \right)^{k} \right| < \dfrac{e}{k} < \dfrac{3}{k}.$

   [This gives an error estimate for the limit in (45.10), one that suggests rather slow convergence.]

References

Bush; Darst; Schaumberger.

C H A P T E R 4 8

POPULATION DYNAMICS: AN INTRODUCTION TO DIFFERENTIAL EQUATIONS

48.1 SIMPLE DIFFERENTIAL EQUATIONS

Back in Chapter 4 we considered perhaps the earliest mathematical treatment of population dynamics, Fibonacci's study of rabbit populations, starting with a single pair of rabbits. The problems of real interest in population dynamics, however, have to do with *large* populations and the effects of overcrowding, competition, limited resources, and so on. When dealing with large populations, even though the size y of the population at any time t is necessarily an integer, it is convenient to *model* the population by treating y as a continuous (indeed, differentiable) function of t and to work from given or supposed information about the *rate* of growth dy/dt. (Decline may be viewed as negative growth.) Suppose, for example, that our given information has the form

$$\frac{dy}{dt} = f(t, y), \qquad y = y_0 \quad \text{when } t = t_0, \tag{48.1}$$

i.e., the rate of growth at any time t is given as an explicit function of the time and the population, and an initial population y_0 is given at the starting time t_0. From such information we would like to be able to predict what the population y will be at any time t, i.e., to find y as an explicit function $g(t)$. Such a function g will have to be a *solution* of (48.1) in the obvious sense:

$$g(t_0) = y_0,$$

and

$$g'(t) = f(t, g(t)) \qquad \text{for } t \geq t_0.$$

The first part of (48.1) is a very special example of a differential equation—i.e., an equation involving a function and one or more of its derivatives. There are usually infinitely many different functions that are solutions of such an equation, but, given side conditions (such as the starting population), we can often reduce the number of solutions to exactly one, at least if the problem has been properly formulated. The technical name for (48.1) is *first-order initial-value problem,* meaning that the differential equation involves only the *first* derivative of y, and the initial population (value of y) has been specified.

The study of differential equations properly belongs to another course beyond the first year of calculus, but it is appropriate to consider its beginnings here because this is really the main reason for studying calculus: Problems from the "real world," whether from biology, chemistry, economics, physics, engineering, or whatever, are likely to be formulated in terms of rates of change, and "solutions" to such problems usually mean functions that satisfy the given derivative conditions. We have selected the particular application to population dynamics only because it suggests (as we will see) several possibilities for the function f in (48.1) and a concrete interpretation of "solution" in terms of prediction of future populations.

The simplest examples of problems of the form (48.1) are those for which dy/dt depends only on t, not on y:

$$\frac{dy}{dt} = f(t), \qquad y = y_0 \quad \text{when } t = t_0. \tag{48.2}$$

In this case a solution is just a particular antiderivative (indefinite integral) of f:

$$y = g(t) = y_0 + \int_{t_0}^{t} f(s)\, ds, \tag{48.3}$$

as you know from the Fundamental Theorem of Calculus. However, even in this case, an expression of the form (48.3) may not tell you anything useful about y if you don't know how to integrate $f(s)$. For example, even if you have spent several weeks studying techniques of formal integration, you still don't know a method for evaluating

$$\int e^{-s^2}\, ds,$$

and for a very good reason: It happens that there is no closed-form expression for this integral in terms of elementary functions. On the other hand, if you write the solution of

$$\frac{dy}{dt} = e^{-t^2}, \qquad y = 0 \quad \text{at } t = 0,$$

as

$$y = \int_{0}^{t} e^{-s^2}\, ds,$$

then, for each fixed value of t, you *can* compute y by using an appropriate numerical integration formula (see Chaps. 24 through 31). Our task for this chapter may be viewed as expanding the idea of numerical integration to include situations like (48.1)—i.e., when the right-hand side of the equation is not given explicitly in terms of t alone.

Another special case of (48.1) has undoubtedly appeared already in your calculus course:

$$\frac{dy}{dt} = ky, \qquad y = y_0 \quad \text{when } t = 0. \tag{48.4}$$

As you know, this is the situation that leads to "exponential growth" (or decay, if $k < 0$), and the unique solution is

$$y = y_0\, e^{kt}. \tag{48.5}$$

This is the simplistic model of population growth that was proposed by the British economist Thomas Malthus (1766–1834), on the basis of which he made dire predictions of future catastrophe, as such rapid (natural) growth could only be held in check by the intervention of war, famine, or some other disaster. The model, of course, neglects too many important characteristics of the way people behave, and it was therefore not very successful for predicting human population growth.

Of course, if you happened to be studying the growth of a bacterial culture in an ideal medium [for which (48.4) is a reasonable model] and you wanted to predict the size of the culture at the end of two days, say, the "solution" (48.5) would still leave you with a computational "problem," which might be solved by a calculator or by the method used in Section 45.3. However, a handy method for proceeding *directly* from (48.4) to numerical values for y is much more efficient than using calculus techniques to obtain an analytic "solution" that leads to another computational problem. Indeed, for the special case of $y_0 = k = 1$, such a method provides another way to compute values of the exponential function. The use of such numerical methods that proceed directly from the differential equation leads one to question which of (48.4) and (48.5) is the "problem" and which is the "solution."

48.2 BIRTH-DEATH MODELS

Before describing a method for solving initial-value problems numerically, let's consider a few more examples of simple population models in situations involving both a birthrate B and a death rate M (for "mortality"). If these rates are both constants, then the population growth rate is given by

$$\frac{dy}{dt} = (B - M)y, \tag{48.6}$$

and we have just the Malthusian model (48.4), with $k = B - M$, the net growth rate. However, it is likely to be the case that B and/or M themselves depend on the size of the population (and other factors as well, which we ignore in the interests of simplicity). For example, if the food supply is limited, the death rate may increase as the population does, due to starvation. If the death rate is proportional to the population, say, $M = my$ for some constant m, then the population growth rate is given by

$$\frac{dy}{dt} = (B - my)y. \tag{48.7}$$

This is the modification of Malthus' model proposed by the Belgian mathematician P. F. Verhulst (1804–1849). Using census data from the period 1790–1840, Verhulst predicted the population of the United States in 1940 and was off by less than 1%. This model has also been used extensively in connection with nonhuman populations, for which the important effects are the natural birth and death rates and competition for available resources.

On the other hand, if there is no limit on resources (or other reason for the death rate to vary), but the birth rate increases proportionally to the population, say, $B = by$, then we have a growth rate of the form

$$\frac{dy}{dt} = (by - M)y. \tag{48.8}$$

It will probably come as no surprise to you to learn that (48.8) is a model for a rather extreme form of "population explosion" (or, in another context, of a nuclear fission process) in which the population becomes infinite in a finite time. Of course, many other models are possible for such birth-death processes by choosing B and M to be other functions of y. As we see later, it is often possible to determine properties of the solutions of such equations (such as inflection points) directly from the equations; and other properties (such as stable limiting populations) may be determined from the analytic solutions. But let us turn now to the problem of computing population as a function of time, i.e., of solving equations of the form (48.1) numerically.

48.3 NUMERICAL SOLUTIONS

We know a starting point (t_0, y_0) on the graph of y, and we may compute the slope of the tangent line at that point from (48.1):

$$\frac{dy}{dt}\bigg|_{t-t_0} = f(t_0, y_0).$$

If $t_1 = t_0 + \Delta t$ is a number close to t_0, then the value of y at t_1 may be approximated by the corresponding y-coordinate on the tangent line (see Fig. 48.1):

$$y_1 = y_0 + f(t_0, y_0)\, \Delta t. \tag{48.9}$$

The point (t_1, y_1) will not lie exactly on the graph of $y = g(t)$, but it will be close. At the true

Figure 48.1

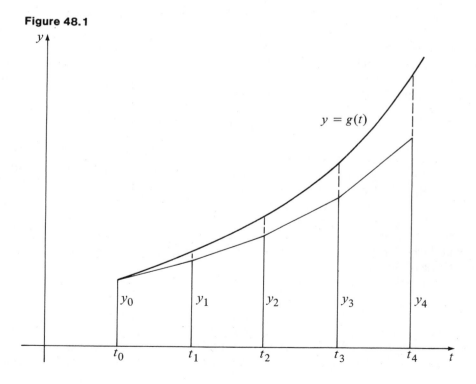

Figure 48.2

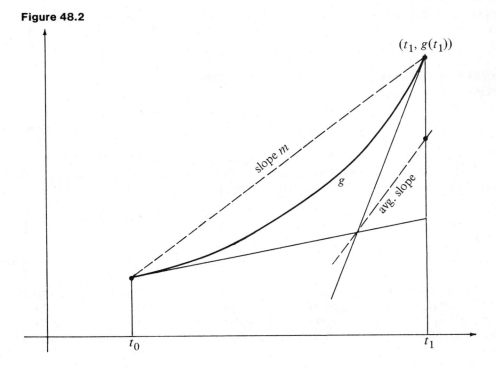

point $(t_1, g(t_1))$ on the graph, the slope would be

$$\frac{dy}{dt}\bigg|_{t=t_1} = f(t_1, g(t_1)).$$

We can't compute that exactly, but we can approximate it by $f(t_1, y_1)$, and then proceed to another approximate value of y at $t_2 = t_1 + \Delta t$:

$$y_2 = y_1 + f(t_1, y_1)\, \Delta t. \tag{48.10}$$

This formula is just like (48.9), and we can obviously repeat this as many times as we wish to compute approximate values $y_1, y_2, y_3, \ldots$, at $t_1, t_2, t_3, \ldots$. The first few steps of this process are illustrated in Figure 48.1.

There is, of course, an obvious flaw in this procedure: The error at each step affects the computations at all succeeding steps, and a small error at first may quickly become magnified into a much larger error. However, *if* Δt is chosen small enough, and *if* the computation is not repeated too many times, and *if* the function f is reasonably "nice," this procedure, crude as it is, will produce acceptable answers to initial-value problems. But we can do better than this without working too much harder, so let's consider a modest improvement in this "tangent method" (also known as the "Euler method").

What we are really trying to approximate (see Fig. 48.2) is the slope

$$m = \frac{g(t_1) - y_0}{\Delta t}$$

of the chord joining the points (t_0, y_0) and $(t_1, g(t_1))$. If we knew m exactly, then we could compute

$$g(t_1) = y_0 + m \, \Delta t.$$

In (48.9) we were approximating m by $f(t_0, y_0)$, but a better approximation could be obtained by averaging the slopes at t_0 and at t_1:

$$m \doteq \frac{1}{2} \left[f(t_0, y_0) + f(t_1, g(t_1)) \right]. \tag{48.11}$$

Of course, there is now a problem in computing the right-hand side of (48.11), as we do not yet know $g(t_1)$; but even a crude approximation [such as that given by the tangent method (48.9)] should give approximately the right slope at t_1. This suggests the following two-step process:

$$z_1 = y_0 + f(t_0, y_0) \, \Delta t; \tag{48.12}$$

$$y_1 = y_0 + \frac{1}{2} \left[f(t_0, y_0) + f(t_1, z_1) \right] \Delta t. \tag{48.13}$$

This is a simple example of what is known as a *predictor-corrector* method. A first, very crude approximation z_1 is "predicted," in this case by the tangent method. Then the crude prediction is "corrected" as in (48.13) by what amounts to the Trapezoidal Rule for numerical integration. Indeed, from (48.1) we could write directly (using the Fundamental Theorem):

$$y_1 = y_0 + \int_{t_0}^{t_1} f(t, g(t)) \, dt;$$

and the trapezoidal approximation to this integral would be

$$\frac{1}{2} \left[f(t_0, y_0) + f(t_1, g(t_1)) \right] \Delta t.$$

Exercise

1. Write a program to tabulate approximate values $y_1, y_2, \ldots, y_{20}$ of $y = g(t)$ defined by (48.1), using formulas (48.12) and (48.13). [Note that y_2 is computed starting from (t_1, y_1)—i.e., by increasing all the subscripts by 1 in the formulas.] The function f may be inserted in the program by a function definition statement, as usual; but note that it will be a function of two variables, unlike those used in earlier programs. Also plan to add a function definition line for $g(t)$, and include $g(t)$ in the output list. The laboratory session begins with problems for which g can be determined analytically, and this permits comparison of the computed approximate values with the true values. For later exercises, g may be deleted. You may take $t_0 = 0$ and $\Delta t = 0.05$.

It should be noted here that we have only scratched the surface of the subject of numerical solution of differential equations. Many more accurate and more powerful methods have been devised. For example, having observed the role of the Trapezoidal Rule in (48.13), it would be natural to ask next what could be done with Simpson's Rule for the integration step, and how we could find *three* points rather than two for making the appropriate curve fit. Furthermore, nothing has been said about the error analysis for such methods, which is left to a later course; we must rely instead on what you are able to observe about magnitudes of errors in your next laboratory session.

48.4 ANALYTIC SOLUTIONS [O]

To finish the discussion of birth-death processes started earlier, let's see what a little calculus can do for us in analyzing equations of the form

$$\frac{dy}{dt} = [B(y) - M(y)]y, \qquad y = y_0 \quad \text{when } t = 0. \tag{48.14}$$

For simplicity, we assume that the birth and mortality rates are linear in y; i.e., $B(y) = a - by$, $M(y) = m + ny$ for some constants a, b, m, n. Then (48.14) reduces to

$$\frac{dy}{dt} = (a - m)y - (b + n)y^2 = \alpha y - \beta y^2, \tag{48.15}$$

where $\alpha = a - m$ and $\beta = b + n$. First, we may ask whether there is a "steady-state" population, i.e., a value of y for which $dy/dt = 0$. If so, by (48.15), we would have

$$0 = y(\alpha - \beta y),$$

or $y = 0$ or α/β. Now $y = 0$ we can rule out as uninteresting. The second solution is possible only if $\beta \neq 0$, of course; but $\beta = 0$ represents the constant birth-death rate case (48.6), about which we already know everything anyway (including the fact that there is no steady-state population). When β and α are both nonzero, we can conclude that a steady-state population is possible if α and β have the same signs, but is impossible otherwise.

Note that when $0 < y < \alpha/\beta$, with β *positive,* then $\alpha - \beta y > 0$, and (48.15) says that y is increasing as a function of t. On the other hand, if $y > \alpha/\beta$ with $\beta > 0$, then $\alpha - \beta y < 0$, and y is decreasing. Thus the population approaches the steady state when $\beta > 0$, regardless of whether it is above or below the steady state. However, as we see later, the steady state is never actually achieved in this case (unless $y_0 = \alpha/\beta$).

Does the graph of y have inflection points, i.e., are there places where the growth rate changes from increasing to decreasing? This can be answered by differentiating (48.15) implicitly with respect to t:

$$\frac{d^2y}{dt^2} = (\alpha - 2\beta y)\frac{dy}{dt}$$
$$= (\alpha - 2\beta y)(\alpha y - \beta y^2)$$
$$= y(\alpha - 2\beta y)(\alpha - \beta y).$$

Now we have ruled out $y = 0$, and $\alpha - \beta y = 0$ only for the steady-state population, so the only possibility for an inflection point is when $\alpha - 2\beta y = 0$, or $y = \alpha/2\beta$—i.e., *half* of the steady state. If the population started out larger than α/β (or, for that matter, larger than $\alpha/2\beta$), this would be impossible. However, a sufficiently small population (with $\alpha, \beta > 0$)

Figure 48.3

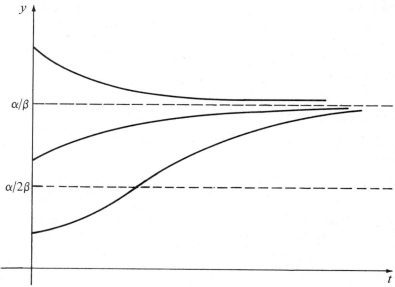

will start out with $d^2y/dt^2 > 0$ (hence with dy/dt increasing); and at half the steady-state population, d^2y/dt^2 will switch to negative values, so dy/dt will be decreasing. Several possibilities are summarized in Figure 48.3, corresponding to different initial populations y_0 at $t = 0$.

Note that all the information so far has been obtained directly from (48.15) without making any attempt to "solve" the differential equation. Now let's observe that we actually can solve it in much the same way that you solved (48.4) in your calculus course to get (48.5). The first step is to rewrite the equation as

$$\frac{dy}{\alpha y - \beta y^2} = dt, \tag{48.16}$$

and then integrate both sides to obtain

$$\frac{1}{\alpha} \ln y - \frac{1}{\alpha} \ln |\alpha - \beta y| = t + C_1. \tag{48.17}$$

Verify. (HINT: Use partial fractions.)

Now if we multiply through by α and take exponentials of both sides, we get

$$\frac{y}{|\alpha - \beta y|} = Ce^{\alpha t}, \tag{48.18}$$

for an appropriate choice of a new constant C. In fact, since $y = y_0$ when $t = 0$, we must have $C = y_0 / |\alpha - \beta y_0|$. For simplicity, let's suppose we have a situation for which $\alpha - \beta y > 0$ (the other possibility may be handled similarly); e.g., the case previously discussed with $y_0 < \alpha/\beta$.

Exercise

2. Solve (48.18) for y, under the preceding assumption, to get

$$y = \frac{\alpha C e^{\alpha t}}{1 + \beta C e^{\alpha t}}. \tag{48.19}$$

Then set $C = y_0/(\alpha - \beta y_0)$ and simplify, to get

$$y = \frac{\alpha y_0}{(\alpha - \beta y_0)e^{-\alpha t} + \beta y_0}. \tag{48.20}$$

Finally, let's make some observations about the solution (48.20) in various cases. If $\beta = 0$, this reduces to (48.5), as it should. When $\beta > 0$, we always have $y < \alpha y_0/\beta y_0 = \alpha/\beta$, which shows that the steady state is never reached. On the other hand, it follows from (48.20) that

$$\lim_{t \to \infty} y = \frac{\alpha}{\beta},$$

so the steady state is approached asymptotically.

In the special case (48.7) (constant birth rate, limited food supply), we have $\alpha = B$, $\beta = m$; so the solution (48.20) becomes

$$y = \frac{B y_0}{(B - m y_0)e^{-Bt} + m y_0}. \tag{48.21}$$

On the other hand, for the "population explosion" case (48.8), we have $\alpha = -M$ and $\beta = -b$; so the solution has the form

$$y = \frac{-M y_0}{(b y_0 - M)e^{Mt} - b y_0}. \tag{48.22}$$

The behavior of the function described by (48.22) depends on the size of y_0 relative to the stable population M/b. If $y_0 > M/b$, then the coefficient of e^{Mt} is positive. As t increases, e^{Mt} also increases, until

$$(b y_0 - M)e^{Mt} = b y_0, \tag{48.23}$$

at which time y becomes infinite! In fact, we can solve (48.23) for the exact time when the explosion takes place (see Fig. 48.4):

$$t_e = \frac{1}{M} \ln \frac{b y_0}{b y_0 - M}. \tag{48.24}$$

However, if $y_0 < M/b$, then we can rewrite (48.22) as

$$y = \frac{M y_0}{(M - b y_0)e^{Mt} + b y_0}.$$

In this case $y \to 0$ as t increases. In fact, (48.8) is a reasonable model for a population that

Figure 48.4

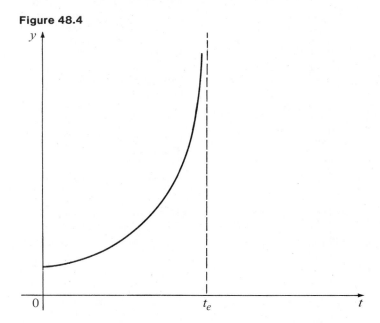

depends on random interactions for breeding. The stable population M/b is then a threshold value, above which the population explodes and below which the species dies out.

References

Bailey (Chap. 1); Bellman and Cooke (Chap. 2); Clark; Defares and Sneddon (Chaps. 9, 10); Maki and Thompson (Chap. 8); Malthus; Pielou (Chap. 1); Simmons (Sect. 1.4, Chap. 2, App. A); D.A. Smith (1977).

C H A P T E R   4 9

PREDICTING POPULATIONS [L]

In each of the following exercises, an initial-value problem in the form

$$\frac{dy}{dt} = f(t, y), \qquad y(0) = y_0$$

is given. You are to use your program from Exercise 1 of Chapter 48 to tabulate the solution on the interval $[0, 1]$. Where it is possible to give an analytic solution $y = g(t)$, we have done so, to be included in your program for comparison purposes. Exercises 4, 5, and 7 are special cases of the growth model (48.15), and the solutions come from (48.20). The solutions to Exercises 2, 3, 6, 8, and 10 may be obtained in a similar manner (see Exercises 17–20), but a method for finding the solution to Exercises 9 and 11 will have to wait for another course. In all cases dealing with population models, you may think it strange to be talking about "populations" with values like $y = 1$, but remember that no units have been specified; $y = 1$ might mean 1000 people, or a million rabbits, or it might be a population *density*, representing one individual per unit of area.

Exercises

1. $dy/dt = y, y(0) = 1$; $g(t) = e^t$. (Program checkout.)

2. $dy/dt = -0.223(y - 21), y(0) = 37$; $g(t) = 21 + 16e^{-0.223t}$.

NOTE: This is a special case of Newton's Law of Cooling, which says that the rate at which a warm body cools in a room at constant temperature is proportional to the temperature differential. The example is taken from the article by Hurley, which contains an interesting application to the case in which the body is that of a murder victim and the temperature at time of death is 37°C, or 98.6°F.

3. $dy/dt = (1 - 0.03y)/1000, y(0) = 0$; $g(t) = (1 - e^{-0.00003t})/0.03$.

NOTE: This model represents the buildup of radioactivity in the vicinity of a nuclear power plant due to the fact that complete containment is not possible. See the article by Clark for further details.

4. $dy/dt = 0.2y - 0.02y^2, y(0) = 1$; $g(t) = 10/(1 + 9e^{-t/5})$.

5. $dy/dt = 0.2y - 0.02y^2, y(0) = 7$; $g(t) = 70/(7 + 3e^{-t/5})$.

NOTE: Exercises 4 and 5 are special cases of the Verhulst population model, one with initial population below, the other above, half the stable population. Notice that $g(t) \rightarrow 10$ as

$t \to \infty$, in both cases. You might find it instructive to graph the solution functions by using the GRAPH program from Chapter 6.

6. $dy/dt = 0.22 + 0.2y - 0.02y^2, y(0) = 1;$ $g(t) = (22e^{0.24t} - 10)/(10 + 2e^{0.24t}).$

NOTE: The Verhulst model, as discussed in Chapter 48, assumed an isolated population. The original version permitted a constant term in the equation to account for a constant net immigration rate. Compare the results here to those for Exercise 4.

7. $dy/dt = 0.2y^2 - 0.1y, y(0) = 10;$ $g(t) = 1/(2 - 1.9e^{t/10}).$

NOTE: This is a population explosion model of the form (48.8).

8. [O] $dy/dt = y^2, y(0) = 0.5;$ $g(t) = 1/(2 - t).$

NOTE: This is an even simpler population explosion model of the form (48.8), with $M = 0$, which invalidates the analytic solution method in Section 48.4. See if you can figure out how $g(t)$ was computed. Except for scaling, this model is equivalent to a "Doomsday" model of world population growth that was proposed in 1960 by Heinz von Foerster and others, only partly with tongues in cheeks. The model was found to fit historical population data better than many more sophisticated models, and it predicted that world population would become infinite on a particularly ominous Friday the 13th in 2026. We can take no comfort in the observation that, more than a generation later, world population is rising *faster* than predicted by the doomsday model (see D. A. Smith [1977] for further details).

9. $dy/dt = -y - ty^2, y(0) = 1;$ $g(t) = 1/(2e^t - t - 1).$

NOTE: It is in the nature of population models that they typically do not involve t in the right-hand side of the equation. However, your program allows the rate of change to depend on t as well as on y. This exercise and the next two are of this sort.

10. $dy/dt = e^{y+t}, y(0) = -1;$ $g(t) = -\ln(e + 1 - e^t).$

11. $dy/dt = -0.179(y + t), y(0) = 30;$ $g(t) = 24.41e^{-0.179t} - t + 5.588.$

NOTE: This exercise is related to Exercise 2. Newton's Law of Cooling can also be applied in situations in which the ambient temperature is *not* constant. In particular, if the murder victim is discovered outdoors, with the ambient temperature at 0°C and dropping at 1°C per hour, a differential equation like the one in this exercise results. The constant 0.179 is an approximate solution to Exercise 19.12. If we set $g(t) = 37$, i.e., normal body temperature, at time of death and try to find t, we arrive at the problem posed by Exercise 19.13 (see D. A. Smith [1978] for further details).

Before proceeding to the next four exercises, delete $g(t)$ from the output statement in your program because no $g(t)$ has been provided for you.

12. $dy/dt = (0.2y - 0.02y^2)/(1 + 0.01y), y(0) = 1.$

NOTE: It was shown in 1963 by F. E. Smith that Verhulst's competition model is not adequate to explain the growth of laboratory colonies of certain water fleas. He proposed a variation of the model of the form shown in Exercise 12 (see Pielou, pp. 30–32, for details).

13. $dy/dt = -y/(y + 1), y(0) = 1.$

NOTE: Problems in chemical kinetics often lead to simple models similar to those in Exercises 2 and 3. However, the work of L. Michaelis and M. Menten in 1913, on enzyme kinetics, led to a differential equation of the form shown in this exercise, for which it is impossible to provide an explicit analytic solution.

14. $dy/dt = 0.2y \ln y - 0.02y^2$, $y(0) = 1$.

NOTE: This exercise and the next are variations of the general population model (48.14). See if you can guess whether the population increases or decreases in each case.

15. $dy/dt = 0.2(y/\ln y) - 0.02y^2$, $y(0) = 2$.

16. [O] You may have felt a little frustrated by the preceding exercises because you were able to compute the solution function only from 0 to 1, and then not very accurately in all cases. Obviously the accuracy could be improved by taking Δt much smaller. But then you would have to wait longer to get answers, even up to $t = 1$. The way around this is to compute far more answers than are printed. Modify your program with a double-loop structure so that it will print n answers, say, but will compute k answers between every pair printed (or nk in all). Then repeat Exercises 4 and 5, taking $n = 10$, $k = 100$; i.e., tabulate the solution for $t = 1, 2, \ldots, 10$, but *compute* with $\Delta t = 0.01$.

17. [O] Solve $dy/dt = a + by$ by "separating the variables" and integrating both sides:

$$\int \frac{dy}{a + by} = \int dt = t + C.$$

Use the result to verify the given solutions in Exercises 2 and 3.

18. [O] Solve the equation in Exercise 8 by separating the variables and integrating.

19. [O] Solve the equation in Exercise 10 by separating the variables and integrating.

20. [O] Solve the equation in Exercise 6 by separating the variables and integrating.

HINT: Use partial fractions.

References

Bellman and Cooke; Clark; Hurley; Kaplan, Ritt, and Tebbs; Pielou; D.A. Smith (1977, 1978).

C H A P T E R 5 0

MORE ON POPULATION DYNAMICS: PREDATORS AND PREY

50.1 FOXES AND RABBITS

Our previous introduction to the subject of differential equations centered around the study of growth or decline of a single population due to changing birth and death rates. The modern mathematical study of ecology includes the study of populations that *interact,* thereby affecting each other's growth rates. In this chapter we consider a very special case of such an interaction, in which there are exactly two species (populations), one of which (the predators) eats the other (the prey). For sake of definiteness, we call the predators "foxes" and the prey "rabbits." For sake of simplicity, we make the following assumptions: The animals live on an island large enough to provide an inexhaustible supply of clover for the rabbits to eat, and there are no outside influences on the two populations. If there were no foxes, the rabbit population would increase at some rate proportional to their number (birth rate minus natural mortality rate). But because there are foxes, there is a negative component of the rate of change of the rabbit population due to being eaten. The foxes, on the other hand, have a limited food supply consisting only of rabbits. If there were no rabbits, the foxes would die off, and we may suppose their rate of decline to be proportional to their number. But there *are* rabbits, and the availability of food permits a positive component of the growth rate for foxes.

Now we come to the crucial assumption. We may suppose that the rate at which foxes and rabbits meet each other is jointly proportional to the sizes of the two populations (i.e., proportional to the product of the populations), and that some fixed proportion of the encounters result in the death of the rabbit. Thus, the negative component of the rabbit growth rate is proportional to the product of the populations. We further assume that what's bad for rabbits is good for foxes, i.e., the positive component of the fox growth rate is also proportional to encounters (or to the product of the populations).

These assumptions may now be summarized in the following system of differential equations, which become our mathematical model for the predator-prey situation. In these equations, x represents the number of rabbits and y the number of foxes, both being functions of time t.

$$\frac{dx}{dt} = ax - bxy; \tag{50.1}$$

$$\frac{dy}{dt} = cxy - py. \tag{50.2}$$

Here a, b, c, and p are constants; ax is the rabbit growth rate in the absence of foxes (i.e., if y were 0, the rabbits would grow in number exponentially), py is the fox death rate in the absence of rabbits, and the other two terms express the decline of rabbits and increase of foxes due to encounters.

Equations (50.1) and (50.2) are called the *Lotka–Volterra Predator-Prey Equations*. They were first formulated in 1920 by the American mathematician-biologist Alfred James Lotka (1880–1949), who used as an example a plant population with unlimited resources for growth and an herbivorous animal population dependent on that plant for food. Vito Volterra (1860–1940) was a famous Italian mathematician who made important contributions to mathematical biology during the last 20 years of his life, after a distinguished career in pure mathematics.

50.2 ANALYSIS OF THE MODEL

What can we expect that a mathematical model like equations (50.1) and (50.2) will predict about the growth and decline of the two populations? Suppose we start with a large number of rabbits (i.e., large x). Then the first term on the right of (50.2) will be the dominant term, and the fox population will increase. When y becomes sufficiently large, the second term on the right of (50.1) will be dominant, and the rabbit population will decrease. But then the food supply for the large number of foxes will dwindle, and foxes will die off more rapidly than they can be replaced. Eventually the smaller fox population should allow the rabbits to make a comeback, and then the cycle might start over again. Is there really this cyclic behavior in both populations, or could a large enough fox population eat all the rabbits, and then die of starvation? To answer such a question, one would like to *solve* the system (50.1) and (50.2) for explicit functions $x = f(t)$, $y = g(t)$, given choices of a, b, c, p, and initial populations x_0, y_0 at $t = 0$. Unfortunately, no method exists for solving such a system of differential equations explicitly. However, it is relatively easy to tabulate approximate values of x and y at $t = t_1, t_2, t_3$, and so on, where $t_i = t_{i-1} + \Delta t$, by essentially the method of Euler (or tangent method) proposed in Chapter 48.

For Δt sufficiently small, we approximate both $x = f(t)$ and $y = g(t)$ in the interval $[0, \Delta t]$ by the respective tangent lines at 0. This leads to approximate values at $t = t_1 = \Delta t$:

$$x_1 = x_0 + \frac{dx}{dt}\bigg|_{t=0} \cdot \Delta t = x_0 + (ax_0 - bx_0y_0)\,\Delta t;$$

$$y_1 = y_0 + \frac{dy}{dt}\bigg|_{t=0} \cdot \Delta t = y_0 + (cx_0y_0 - py_0)\,\Delta t.$$

Then we proceed to approximate values at $t = t_2 = 2 \cdot \Delta t$ by making another tangent-line approximation, assuming that the slopes are given correctly by (50.1) and (50.2) for $x = x_1$, $y = y_1$:

$$x_2 = x_1 + (ax_1 - bx_1y_1)\,\Delta t;$$
$$y_2 = y_1 + (cx_1y_1 - py_1)\,\Delta t.$$

This process may be continued indefinitely, but, of course, the approximations are crude unless Δt is quite small, and get worse as time goes on in any case. However, the "predicted" values for each new x and y may be "corrected," just as they were in Chapter 48, by a trapezoidal corrector. The formulas for doing this are a little easier to express if we introduce functional notation for the right-hand members of equations (50.1) and (50.2):

$$F(x, y) = ax - bxy;$$ (50.3)

$$G(x, y) = cxy - py.$$ (50.4)

These functions express the *time rates of change* of the populations x and y as functions of the populations themselves. (In principle, these rates might depend on t too, but our assumptions about the interactions were independent of time.) Now all we have to do is replicate the predictor-corrector formulas (48.12) and (48.13), remembering that we are dealing with two *dependent* variables, and using F and G for the rates of change at the appropriate points, namely, the left- and right-hand endpoints of each time interval. To get from step 0 to step 1, we have

$$w_1 = x_0 + F(x_0, y_0)\, \Delta t;$$ (50.5)

$$z_1 = y_0 + G(x_0, y_0)\, \Delta t;$$ (50.6)

$$x_1 = x_0 + \frac{1}{2}\,[F(x_0, y_0) + F(w_1, z_1)]\, \Delta t;$$ (50.7)

$$y_1 = y_0 + \frac{1}{2}\,[G(x_0, y_0) + G(w_1, z_1)]\, \Delta t.$$ (50.8)

As before, we move on to time t_2 by increasing all the subscripts by 1, and so on.

Because this problem is already rather complicated, you will not be asked to write a program this time, but rather to use one already prepared for you that has been saved under the name FOXRAB.[1] Instructions for using this program are provided at the start of the next laboratory session. The remainder of this chapter is devoted to a discussion of consequences of the predator-prey model that can be obtained directly from equations (50.1) and (50.2).

50.3 TRAJECTORIES

Our unknown solution functions $x = f(t)$, $y = g(t)$, which give the two populations at time t, may be thought of as parametric equations for a curve in the x, y-plane. Such a curve is called a *trajectory* of the system, and there happens to be a unique trajectory through each possible starting point (x_0, y_0). Each such trajectory is also a solution of an initial-value problem that arises by computing dy/dx from (50.1) and (50.2):

$$\frac{dy}{dx} = \frac{dy/dt}{dx/dt} = \frac{cxy - py}{ax - bxy}, \qquad y = y_0 \quad \text{when } x = x_0.$$ (50.9)

(We ignore points at which the denominator is 0 for the time being.) In form, (50.9) is just like the initial-value problems (48.1) considered earlier, with x playing the role of independent variable, instead of t. Indeed, (50.9) can be "solved" by the separation-

---

[1] This program has been adapted from one prepared by the Biology Curriculum Group, University of Illinois, Chicago Circle Campus. Complete listings are given in the *Instructor's Manual*.

of-variables technique:

$$\frac{dy}{dx} = \frac{(cx - p)y}{(a - by)x},$$

$$\frac{a - by}{y} dy = \frac{cx - p}{x} dx,$$

$$\int \left(\frac{a}{y} - b\right) dy = \int \left(c - \frac{p}{x}\right) dx,$$

$$a \ln y - by = cx - p \ln x + C.$$

Taking exponentials of both sides and introducing a new version of the constant of integration leads to the following equation for a trajectory:

$$x^p y^a e^{-(cx + by)} = k, \tag{50.10}$$

where k is the constant $x_0^p y_0^a e^{-(cx_0 + by_0)}$. However, (50.10) cannot be solved explicitly for either x or y, and none of the properties of the trajectories are particularly apparent from this equation, except one: If x_0 and y_0 are both positive, then $k > 0$, so neither x nor y can ever take the value 0, i.e., neither population ever dies off completely. [On the other hand, both x and y can come arbitrarily close to 0 if k is sufficiently small. A "realistic" interpretation of the model would have to include the observation that if the rabbit population ever drops below 2, say, then (50.1) no longer applies.]

There is a clever device for graphing equations of the form (50.10), due to Volterra, that is based on examination of the graphs of $x^p e^{-cx}$ and $y^a e^{-by}$. (For a good exposition of this method, see Simmons, pp. 284–288.) Notice that it follows from (50.1) and (50.2) that $x_0 = p/c$, $y_0 = a/b$ are *stable* populations, in that both dx/dt and dy/dt are 0 for these starting values, and hence remain 0. For initial populations that are near these stable values, it can be shown that the trajectories are nearly elliptical, with the stable point $(p/c, a/b)$ as center. Farther away from the stable point, the shapes of the curves become less elliptical, but they remain *closed* curves, i.e., after a certain period of time, the same pairs (x, y) are repeated over again. Several of these trajectories (for $a = 4$, $b = 2$, $c = 1$, $p = 3$) are shown in Figure 50.1, corresponding to $k = 0.03, 0.18, 0.285, 0.38$, starting with the outermost curve.

The trajectories may now be interpreted to draw the following conclusions about the model (50.1) and (50.2). The populations of rabbits and foxes go through four phases, represented by the Roman numerals in Figure 50.1: If we start anywhere in region I, we have an abundance of rabbits ($x > p/c$), and the fox population is increasing, but the rabbit population is decreasing. When x drops to p/c, the relative scarcity of food causes the fox population to start to decline (region II). There are still enough foxes ($y > a/b$), however, that the rabbit population is still declining. When y drops to a/b, there are few enough foxes that the rabbits start to make a comeback (region III). When their numbers reach p/c again, the food supply for foxes becomes abundant enough for them to make a comeback also (region IV), and both populations increase until $y = a/b$ again, at which time we reenter region I, and the whole process repeats itself.

For a more detailed exposition of the predator-prey model, see Chapter 3 of Kemeny and Snell. That chapter also includes a discussion of a somewhat similar model for a situation in

Figure 50.1

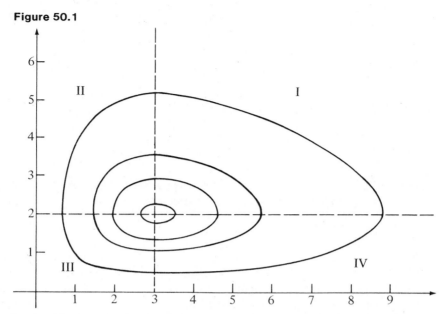

Reprinted from *Mathematical Models in the Social Sciences* by J.G. Kenney and J.L. Snell, by permission of the M.I.T. Press, Cambridge, Massachusetts. Copyright © 1962 by Ginn and Company.

which each of two species can kill the other. In spite of the similarity of the models, the results are strikingly different in this case (see Exercise 51.4).

References

Bailey (Sect. 8.4); Kaplan, Ritt, and Tebbs; Kemeny and Snell (Chap. 3); Lotka (1920; 1925, Part 2); Maki and Thompson (Chap. 8, Part A); Pielou (Part I); Rescigno and Richardson; Simmons (Chap. 7).

CHAPTER 51

CAN RABBITS LIVE WITH FOXES? [L]

The exercises in this laboratory session are to be done with the foxes-and-rabbits program FOXRAB, which we now proceed to describe in detail. As was the case with GRAPH in Chapter 6, there is a "low-resolution" version FOXRAB.LR that will run on virtually any BASIC-speaking computer (preferably with printed output) and a "high-resolution" graphic version FOXRAB.HR for supported microcomputers. Both versions have the following features:

Numerical solutions of the differential equations (50.1) and (50.2) are computed by the predictor-corrector formulas (50.5) to (50.8) over the time interval [0, 5], where time is expressed in years. This interval is divided into 100 subintervals, so $\Delta t = 0.05$. The initial populations $X0$ and $Y0$ and the growth and death factors A, B, C, and P are input variables. The variables X and Y represent population sizes for rabbits and foxes, respectively, in hundreds. Permissible values for all six input variables are positive numbers less than ten.

FOXRAB.LR prints a tabulation of the populations at every fifth time step (i.e., at quarter years) and character graphic representations of x and y as functions of t. If you don't have a printer (or choose not to use it), the tabulation can be printed on a video screen. However, a 40-column screen is not wide enough to accommodate the graphs.

FOXRAB.HR offers three options for output, plus the ability to switch among them as often as you wish after each solution is computed: (1) a numerical tabulation as in FOXRAB.LR; (2) a high-resolution graph of x and y as functions of t; and (3) a high-resolution graph of the trajectory (50.10), similar to Figure 50.1. Option (3) demonstrates more vividly than (1) or (2) the cyclic nature of the solutions.

Exercises

1. Run FOXRAB with $A = 4$, $B = 2$, $C = 1$, $P = 3$, and the following pairs of starting populations: (a) $X = 3$, $Y = 0.5$; (b) $X = 3$, $Y = 1$; (c) $X = 3$, $Y = 1.3$; (d) $X = 3$, $Y = 1.75$; (e) $X = 3$, $Y = 2$.

NOTE: These values of the parameters correspond to the trajectories shown in Figure 50.1.

2. Run FOXRAB with $A = 4$, $B = 4$, $C = 0.8$, $P = 2$, and the following pairs of starting populations: (a) $X = 4$, $Y = 2$; (b) $X = 2.5$, $Y = 2.5$; (c) $X = 2.5$, $Y = 1.5$; (d) $X = 2.8$, $Y = 1$; (e) $X = 2.5$, $Y = 1$.

3. Suppose our island has an apparent abundance of rabbits, which attracts rabbit hunters who come to the island on a regular basis. The rate at which they kill rabbits will be proportional to the rabbit population, and their effect will be seen in smaller values for the parameter A, representing the natural growth rate for rabbits. In all parts of this problem, take the initial populations to be $X = 8$, $Y = 1$, and let $B = 1$, $C = 0.7$, and $P = 2$.

a. Suppose first that the toll taken by rabbit hunters exactly offsets the birth rate, so that $A = 0$. Run FOXRAB to see what happens. (The result may be a bit of a surprise.)

b. It is unlikely, of course, that the hunters could exactly match the birth rate with their kill. To see what happens if they kill too many rabbits, run FOXRAB with $A = -1$. (This result will be less surprising, except that the foxes starve before all the rabbits are gone.)

c. If the effect of the hunters is to reduce A to a small positive value, then we still have the Lotka–Volterra situation, in which there should be cyclic variations of the populations. However, the results can still be precarious for the foxes, as you can check by running FOXRAB with $A = 0.5$.

d. To see that recovery for the foxes is possible, repeat part (c) with $A = 0.9$.

NOTE: It will be clear from the results of this problem that FOXRAB does not require that the inputs be positive, even though that is required by the Lotka–Volterra model. This observation is relevant for the next problem as well.

4. The population model for competing species, each of which can kill the other, referred to in the last paragraph of Chapter 50, has the following form:

$$\frac{dx}{dt} = ax - bxy, \tag{51.1}$$

$$\frac{dy}{dt} = py - cxy, \tag{51.2}$$

where a, b, c, and p are positive. This represents a natural growth rate for each species, modified by the negative effect of encounters. This model can be simulated by FOXRAB by entering negative values for the parameters C and P. Run FOXRAB with $A = 1$, $B = 0.5$, $C = -0.25$, $P = -0.75$, and the following pairs of starting populations: (a) $X = 1$, $Y = 0.5$; (b) $X = 0.5$, $Y = 1$; (c) $X = 6$, $Y = 3$; (d) $X = 6$, $Y = 4$.

NOTE: The output is still labeled "RABBITS" and "FOXES," of course. Imagine that the rabbits have armed themselves to fight back, and both species have learned to eat either clover or each other.

The remaining (optional) exercises provide a glimpse of other models of interacting populations. In order to do these exercises, you have to modify the FOXRAB program. Both versions have the rate-of-change functions F and G defined on lines 500 and 510 (via DEF FNF and DEF FNG, respectively). By changing these definitions, you can use FOXRAB to solve other systems of differential equations. Exercises 6 and 8 also require two new coefficients, Q and R. These may be inserted in the programs by assignment or input statements, and they will not conflict with variable names already in the programs.

Exercises

5. [O] The branch of mathematics known as Operations Research is often described as having started with mathematical studies of military operations during World War II.

However, during World War I a British mathematician, F. W. Lanchester, proposed the following model for the sizes of competing armies:

$$\frac{dx}{dt} = -ay, \tag{51.3}$$

$$\frac{dy}{dt} = -bx, \tag{51.4}$$

where a and b represent the firepower effectiveness of the y and x forces, respectively. Solve the Lanchester equations for various choices of parameters (on the order of 0.1 to 0.5) and various sizes of initial forces.

NOTE: Equations (51.3) and (51.4) are really expected to hold only as long as both populations are positive. If your solution proceeds past the point where one force is exterminated, it will demonstrate behavior quite unlike armies.

6. [O] Many variations of the predator-prey and competing-species model have been proposed and studied. For example, a more realistic model than the Lotka–Volterra equations might be represented by

$$\frac{dx}{dt} = ax - bx^2 - cxy, \tag{51.5}$$

$$\frac{dy}{dt} = py - qy^2 + rxy. \tag{51.6}$$

The terms in x^2 and y^2 represent competition within each species for available resources (clover is not unlimited!), and the term in y allows for a natural growth rate for the predators. (Foxes may have a food supply other than rabbits.) Exploration of this model reveals quite different behavior from the cyclic variations of Exercises 1 and 2. Suggested starting values: $X = 6$, $Y = 1$; $A = 2$, $B = 0.25$, $C = 0.5$, $P = 1$, $Q = 0.5$, $R = 0.5$. Also try other combinations of X and Y with the same parameters.

7. [O] Another predator-prey model (see Pielou, Sect. 6.3) has the following form:

$$\frac{dx}{dt} = ax - bxy, \tag{51.7}$$

$$\frac{dy}{dt} = py - \frac{cy^2}{x} = \left(p - \frac{cy}{x} \right) y. \tag{51.8}$$

The prey equation (51.7) is the same as in the Lotka–Volterra case. The predator equation allows a natural growth rate p and a competition term based on the ratio y/x. When this ratio is large, the growth of predators slows down. However, when y/x is small (not too many foxes and an abundance of rabbits), the predators grow at nearly their natural growth rate. Explore this model with an appropriate modification of FOXRAB. Suggested starting values: $X = 8$, $Y = 2$; $A = 2$, $B = 2$, $C = 5$, $P = 2$. Try other combinations as well. Also try adding a term in x^2 to (51.7), to make it look more like (51.5); i.e., limit the clover again.

8. [O] The competition model (51.1) and (51.2) of Exercise 4 could be made more realistic in some cases by adding terms for intraspecies competition, as in Exercise 6:

$$\frac{dx}{dt} = ax - bx^2 - cxy, \tag{51.9}$$

$$\frac{dy}{dt} = py - qy^2 - rxy. \tag{51.10}$$

(The only difference between this and the model in Exercise 6 is that interaction is bad for both species.) This model is analyzed in detail by Pielou, Chapter 5, where she shows that various combinations of the parameters and initial populations lead to one or the other of the populations dying out, or both approaching stable sizes. Suggested starting values for your exploration:

a. $X = 2$, $Y = 3.5$; $A = 4$, $B = 0.28$, $C = 0.4$, $P = 3$, $Q = R = 0.28$.
b. $X = 5$, $Y = 2$; $A = 10$, $B = 1.4$, $C = 1.2$, $P = 8$, $Q = 1$, $R = 0.9$.

References

Kemeny and Snell (Chap. 3); Lanchester; Pielou (Part I); Rescigno and Richardson; Stoller (Chap. 1).

C H A P T E R 5 2

POSITIVE-TERM SERIES AND THE INTEGRAL TEST

On several earlier occasions (Chaps. 4, 12, 37, and 44) we hinted at or openly discussed the possibility of adding up an infinite number of terms. This means, of course, computing a limit of a sequence of "partial sums," each of which is just an ordinary sum of a finite number of terms. To use a computer to calculate such a limit, we use calculus to determine when a partial sum is sufficiently close to the limit, and compute that partial sum instead. In this and several succeeding chapters, we explore how the ideas usually presented in the calculus course for determining whether a given series converges can be used to aid in computing the value of a series when it does converge. (Note that the exercises in the series chapter in your calculus book ask you to find the values of very few infinite series.)

52.1 AN IMPROVEMENT ON THE INTEGRAL TEST

The integral test for determining convergence of a series $\Sigma_{k-1}^{\infty} a_k$ is particularly useful when $a_k = f(k)$ for some continuous, nonnegative function f for which an antiderivative is easily found. The test is simply that the series converges if and only if $\int_1^{\infty} f(x)\, dx$ converges. Let us recall how this is established.

In Figure 52.1 we have a portion of the graph of f and some horizontal lines through the points (n, a_n), $(n + 1, a_{n+1})$, and so on. (Ignore the diagonal lines for the moment.) Looking at the interval $[n, n + 1]$, we see that

$$a_{n+1} \le \int_n^{n+1} f(x)\, dx \le a_n, \qquad (52.1)$$

since the first and last members are areas of inscribed and circumscribed rectangles, and the middle member is the area under the graph of f. Summing the inequalities (52.1) from $n = 1$ to ∞ (actually, taking limits of partial sums) leads to

$$\sum_{k-2}^{\infty} a_k \le \int_1^{\infty} f(x)\, dx \le \sum_{n-1}^{\infty} a_n, \qquad (52.2)$$

and hence to the conclusion that the series converges if and only if the integral does.

Figure 52.1

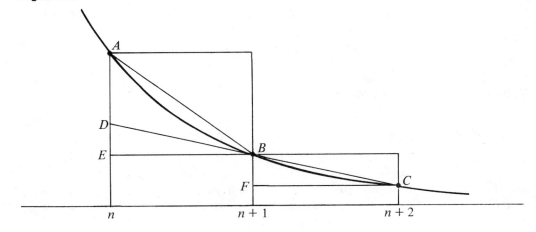

Now the rectangular areas in (52.1) are actually rather crude approximations to the integral, and we can get much better ones by choosing appropriate linear approximations to f. (Recall the distinction between calculating integrals by Riemann sums and by the Trapezoidal Rule.) As it is the series, however, and not the integral we are trying to compute, it is important that the trapezoidal approximating areas be closely related to terms of the series. Refer again to Figure 52.1, where we now assume $f'(x) < 0$, $f''(x) > 0$, and $\int_1^{\infty} f(x)\, dx$ exists, so the series actually has a sum. The line segments AB and BC are chords of the graph of f, and BD is an extension of BC back to $x = n$. The area of triangle ABE is

$\frac{1}{2} (a_n - a_{n+1})$, and that of $\triangle BCF$ is $\frac{1}{2} (a_{n+1} - a_{n+2})$. (Why?) Note that $\triangle DBE$ is congruent to $\triangle BCF$. (Why?) Hence it has the same area. Since the rectangle below segment EB has area a_{n+1}, consideration of the appropriate trapezoidal areas shows that

$$\frac{3}{2} a_{n+1} - \frac{1}{2} a_{n+2} \le \int_n^{n+1} f(x) \, dx \le \frac{1}{2} a_n + \frac{1}{2} a_{n+1}. \tag{52.3}$$

(Why?) Next we add $a_{n+2}/2$ to each member of (52.3) and subtract $a_{n+1}/2$:

$$a_{n+1} \le \int_n^{n+1} f(x) \, dx - \frac{1}{2} a_{n+1} + \frac{1}{2} a_{n+2} \le \frac{1}{2} (a_n + a_{n+2}). \tag{52.4}$$

Each member of (52.4) may now be summed from $n = N$ to ∞. We start on the right:

$$\sum_{n=N}^{\infty} \frac{1}{2} (a_n + a_{n+2}) = \frac{1}{2} \sum_{n=N}^{\infty} a_n + \frac{1}{2} \sum_{k=N+2}^{\infty} a_k$$

$$= \sum_{k=N+1}^{\infty} a_k + \frac{1}{2} a_N - \frac{1}{2} a_{N+1}. \qquad \text{(Why?)}$$

For the middle member, the sum is

$$\int_N^{\infty} f(x) \, dx - \frac{1}{2} \sum_{k=N+1}^{\infty} a_k + \frac{1}{2} \sum_{k=N+2}^{\infty} a_k = \int_N^{\infty} f(x) \, dx - \frac{1}{2} a_{N+1}. \qquad \text{(Why?)}$$

The sum for the left-hand member of (52.4) is obvious, so the result of summing (52.4) is

$$\sum_{k=N+1}^{\infty} a_k \le \int_N^{\infty} f(x) \, dx - \frac{1}{2} a_{N+1} \le \sum_{k=N+1}^{\infty} a_k + \frac{1}{2} (a_N - a_{N+1}). \tag{52.5}$$

Notice that the sum that appears twice in (52.5) is just the remainder after summing the first N terms of the series. If we add that partial sum to each member of (52.5), we get

$$\sum_{k=1}^{\infty} a_k \le \sum_{k=1}^{N} a_k + \int_N^{\infty} f(x) \, dx - \frac{1}{2} a_{N+1} \le \sum_{k=1}^{\infty} a_k + \frac{1}{2} (a_N - a_{N+1}). \tag{52.6}$$

Notice that the left and right members of (52.6) differ by $\frac{1}{2}(a_N - a_{N+1})$, a quantity we should expect to be small when N is large. Indeed, (52.6) tells us that if we approximate the sum of the series by the middle member, the error will not exceed $\frac{1}{2}(a_N - a_{N+1})$. This is an approximation by a "corrected" partial sum: After summing N terms, we correct by adding $\int_N^{\infty} f(x) \, dx$ (to be computed by using calculus) and subtracting $a_{N+1}/2$.

52.2 EXAMPLE: p-SERIES

Among those series for which the previously described method is applicable are the important family of p-series: $\sum_1^{\infty} n^{-p}$. As you know from the ordinary integral test, such a series is convergent if and only if $p > 1$. The integral that appears in (52.6) is

$$\int_N^{\infty} \frac{1}{x^p} \, dx = \frac{1}{(p-1)N^{p-1}}.$$

Figure 52.2

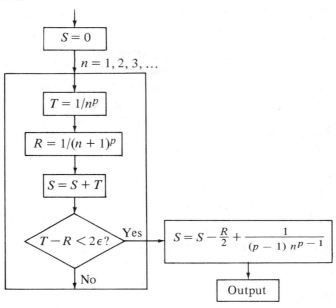

(Verify!) A partial flowchart for the evaluation of a p-series by using this method is shown in Figure 52.2. Note that terms of the sum are accumulated until the end test is satisfied, and then the correction term is added.

Exercises

1. Complete the flowchart and write a program to evaluate the sum of a p-series, for any $p > 1$, to five decimal places. Output should include the number N of terms used.

2. For $p = 1$, the p-series is the harmonic series $1 + \frac{1}{2} + \frac{1}{3} + \cdots$, known to be divergent. Nevertheless, if you asked a computer to add up its partial sums, eventually you would get a constant value for all successive sums. Can you think why? (The constant value obtained would depend on the computer used, but a typical result would be to get a sum of a little over 18 when N is on the order of 10^8.) This example and one we consider in the next laboratory session illustrate the need to determine whether or not the series converges before using the computer.

52.3 EXAMPLE: THE ALTERNATING HARMONIC SERIES [O]

The method represented by inequalities (52.6) applies only to positive-term series because of the area considerations in its derivation. Nevertheless, it may be applied to some alternating series (see Chap. 54) by means of the following device: If $b_1 - b_2 + b_3 - b_4 + \cdots$ is an alternating series (i.e., $b_n > 0$ for each n) that is known to converge, then the terms may be grouped as

$$(b_1 - b_2) + (b_3 - b_4) + \cdots$$

and each pair of terms replaced by a term of a new series, say, $a_n = b_{2n-1} - b_{2n}$, $n = 1, 2, \ldots$. In particular, if the b's are strictly decreasing, then the a's will all be positive. Partial sums of the a's are just the even-numbered partial sums of the b's, and must therefore have the same limit. If the method of this chapter applies to the sum of the a's, it will usually give a much sharper error estimate than the alternating-series test gives.

We illustrate this in the case of the alternating harmonic series $1 - \frac{1}{2} + \frac{1}{3} - \frac{1}{4} + \cdots$. Thus, $b_n = 1/n$, and

$$a_n = \frac{1}{2n - 1} - \frac{1}{2n} = \frac{1}{2n(2n - 1)}. \tag{52.7}$$

The integral needed for the correction is

$$\int_N^\infty \frac{dx}{2x(2x - 1)} = \frac{1}{2} \ln\left(1 + \frac{1}{2N - 1}\right).$$

Verify!

HINT: The partial-fraction decomposition of a_n is given in (52.7).
The error estimate needed for deciding when to stop accumulating the sum is

$$\frac{1}{2}(a_n - a_{n-1}) = \frac{4n + 1}{(2n - 1)(2n)(2n + 1)(2n + 2)}. \tag{52.8}$$

Verify.

Exercises

3. Carry out the verifications previously indicated. Then show that the right-hand member of (52.8) is $< \frac{1}{4}n^3$. Use of the latter for the error estimate simplifies programming somewhat.

HINT: First increase the numerator in (52.8) by one to get a simpler, but slightly larger, expression.

4. Write a program to evaluate $\sum_{n=1}^\infty (-1)^{n-1}/n$ by the method previously described.

5. [O] Carry out the analysis required for applying the method of this chapter to evaluate $\sum_{n=1}^\infty (-1)^{n-1}/n^2$.

References

Ayoub; Boas; Gardner (1964); Morley; Porter.

CHAPTER 53

LET'S COMPUTE p - SERIES [L]

Exercises

1. Use your program from Exercise 1 of Chapter 52 to evaluate the p-series corresponding to the following values of p: (a) 2; (b) 2.5; (c) 3; (d) 4; (e) 7.

NOTE: Part (a) is the Euler sum discussed in Chapter 4. Recall that the answer is $\pi^2/6$, which you can check directly.

2. Run your program again with $p = 0.99$. What is the significance of your answer?

3. [O] Run your program for Exercise 4 of Chapter 52 to evaluate the sum of the alternating harmonic series.

NOTE: The correct answer is ln 2, the value of which you can obtain by other means.

CHAPTER 54

THE UPS AND DOWNS OF ALTERNATING SERIES

Your calculus book contains a discussion of alternating series that shows that there is a very easy test for convergence. Specifically, if the series has the form

$$a_1 - a_2 + a_3 - a_4 + \cdots \tag{54.1}$$

with $a_n > 0$ for each n, and if the a's decrease to zero (i.e., if $a_n \geq a_{n+1}$ for each n, and $\lim_{n \to \infty} a_n = 0$), then the series (54.1) converges. Furthermore, the argument used to prove this provides an error estimate: The remainder after summing N terms is not more than a_{N+1}. Unfortunately, that error estimate is not too useful if the a's converge rather slowly to zero. For example, if $a_n = 1/n$ (the alternating harmonic series), to get four-place accuracy in the answer, you would have to add 20,000 terms of the sum. (Why?) By contrast, the device of pairing terms and using a correction term, as sketched in Section 52.3, will assure four-place accuracy with only 14 terms of the paired series, or 28 of the original series. (Verify, using

Exercise 3 of Chap. 52.) However, if you attempted Exercise 5 of Chapter 52, you already know that it is not very easy to apply the method of that chapter to alternating series in general. In this chapter we describe a method for "correcting" a partial sum of a more or less arbitrary alternating series with an error estimate that is easy to compute, and much more reasonable than the "next-term" estimate of the alternating-series test.

In addition to assuming the conditions of the alternating-series test, we assume that successive differences of the a's are also decreasing, i.e.,

$$a_n - a_{n+1} \geq a_{n+1} - a_{n+2}, \qquad n = 1, 2, 3, \ldots . \tag{54.2}$$

Our estimate of the remainder after summing N terms of (54.1) is obtained by "splitting" all the remaining terms in half and taking half the $(N + 1)$-th term as the "correction" term:

$$\sum_{n=N+1}^{\infty} (-1)^{n-1} a_n = (-1)^N \left[a_{N+1} - a_{N+2} + a_{N+3} - \cdots \right]$$

$$= (-1)^N \left[\frac{1}{2} a_{N+1} + \frac{1}{2} (a_{N+1} - a_{N+2}) \right.$$

$$\left. - \frac{1}{2} (a_{N+2} - a_{N+3}) + \frac{1}{2} (a_{N+3} - a_{N+4}) - \cdots \right]$$

$$= \frac{(-1)^N}{2} a_{N+1} + \frac{1}{2} \sum_{n=N+1}^{\infty} (-1)^{n-1} (a_n - a_{n+1}).$$

The last sum is another alternating series, since $a_n \geq a_{n+1}$, and its terms decrease to 0 in absolute value, by (54.2) and the fact that $\lim_{n \to \infty} a_n = 0$. Hence, by the usual version of the alternating-series test, that sum is smaller in absolute value than its first term, namely, $\frac{1}{2}(a_{N+1} - a_{N+2})$. It follows that

$$\left| \sum_{n=N+1}^{\infty} (-1)^{n-1} a_n - \frac{(-1)^N}{2} a_{N+1} \right| < \frac{1}{2} (a_{N+1} - a_{N+2}). \tag{54.3}$$

Now if we add up N terms of (54.1), and "correct" the sum by adding half the next term [i.e., approximate the remainder as indicated in (54.3)], the resulting error is not more than the right-hand member of (54.3).

NOTE: The technique of correcting the partial sum by adding half the next term is the simplest special case of a family of methods for evaluating alternating series known collectively as the *Euler transform*. See Johnsonbaugh for a more detailed analysis of this and related methods. Both here and in Chapter 52 we have effectively replaced the sequence of partial sums of a given series by another sequence ("corrected partial sums") that converges faster than the original. There are many such methods for evaluating limits of sequences and series; the technical term for such methods is *acceleration of convergence*. Another popular acceleration method is introduced in Chapter 56.

A partial flowchart for this algorithm is shown in Figure 54.1. The absolute value of the nth term, a_n, may be defined by a function definition statement to make the program as general as possible. Note the use of the sign factor q (alternately 1 and -1) to keep track of the alternating signs. As in Figure 52.2, the sum is accumulated until half the difference of the next two terms is small enough, and then the correction term is added.

Figure 54.1

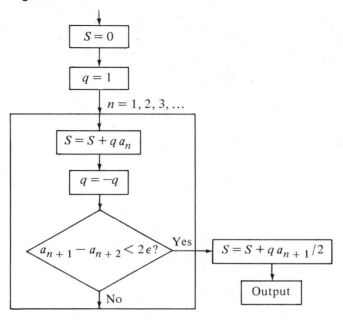

Exercises

1. Complete the flowchart of Figure 54.1 and write a program to implement it. Take $\epsilon = 0.00005$. Your output should include the number of terms required in the sum.

2. Verify that condition (54.2) is satisfied for
   a. $a_n = 1/n$
   b. $a_n = 1/n^2$

3. Verify condition (54.2) for each of the following:
   a. $a_n = 1/(2n - 1)$
   b. $a_n = 1/2n(2n - 1)$

4. Suppose $a_n = f(n)$, where $f(x)$ is a function defined on $[1, \infty)$ such that $f'(x) < 0, f''(x) > 0$ for all x. Show that condition (54.2) is satisfied. (HINT: Use the Mean Value Theorem.)

5. Show that (54.2) is satisfied for any alternating p-series, where p is a positive rational number. (HINT: Use Exercise 4.)

References

Henriksen and Lees (Sect. 28, Exercises 20–23; App. A, Problem 28.4); Johnsonbaugh; Pinsky.

CHAPTER 55

LET'S EVALUATE ALTERNATING SERIES [L]

The following exercises present alternating series to be evaluated by your program from Exercise 1 of Chapter 54, which should produce answers that are correct to four places. In each case it is very easy to see that the series converges. However, it is not obvious in most cases that condition (54.2) is satisfied, a condition that is necessary in order for your program to be meaningful. Exercises 2 through 5 of Chapter 54 cover all but the last three following cases.

Exercises

1. $\Sigma_{n-1}^{\infty} (-1)^{n-1}/n^p$, for the following values of p: (a) 1; (b) 2; (c) 2.5; (d) 4; (e) 7; (f) 0.9.

2. $\Sigma_{n-1}^{\infty} (-1)^{n-1}/(2n - 1)$. (This is the Leibniz series, discussed in Chap. 12, the value of which is $\pi/4$.)

NOTE: The following four series were evaluated analytically 60 years ago (see Underwood). The answers in parentheses are the results of those evaluations.

3. $\Sigma_{n-1}^{\infty} \dfrac{(-1)^{n-1}}{2n(2n - 1)}$ (*Answer:* $\dfrac{\pi}{4} - \dfrac{1}{2} \ln 2$.)

4. $\Sigma_{n-1}^{\infty} \dfrac{(-1)^{n-1}}{(2n + 1)(2n)(2n - 1)}$ (*Answer:* $\dfrac{1}{2} - \dfrac{1}{2} \ln 2$.)

5. $\Sigma_{n-1}^{\infty} \dfrac{(-1)^{n-1}}{(2n + 2)(2n + 1)(2n)(2n - 1)}$ (*Answer:* $\dfrac{5}{12} - \dfrac{\pi}{12} - \dfrac{1}{6} \ln 2$.)

6. $\Sigma_{n-1}^{\infty} \dfrac{(-1)^{n-1}}{(2n + 3)(2n + 2)(2n + 1)(2n)(2n - 1)}$ (*Answer:* $\dfrac{5}{36} - \dfrac{\pi}{24}$.)

7. As noted earlier, condition (54.2) has not been verified for Exercises 4, 5, and 6. However, your program should have produced answers, and they can be checked against answers obtained another way. Do it.

8. Verify condition (54.2) for the series in Exercise 4. (HINT: Use the result of Exercise 4 of Chap. 54.)

CHAPTER 56

WHEN RATIOS DECREASE TO A LIMIT LESS THAN ONE

56.1 THE RATIO TEST

One of the most popular tests for convergence of a series $\Sigma\, a_n$ is the ratio test, which says that if $\lim_{n \to \infty} |a_{n+1}/a_n| < 1$, then the series converges. This test is particularly useful when the terms a_n involve factorials and in some other situations in which the ratio of terms is a simpler object of study than the term itself. There are many series, however, for which the ratio test gives no information (all the p-series, e.g.) by virtue of having ratios that approach 1.

Recall how the ratio test shows convergence when the limit is less than 1: This fact allows comparison with a geometric series with common positive ratio $r < 1$, and every such series converges (Theorem 11.2, or see your calculus book). The same sort of comparison allows us to estimate the remainder after summing N terms, as we now see.

Suppose, for convenience, that the a_n's are all positive. (If not, we can work with their absolute values because absolute convergence implies convergence.) Suppose further that

$$\frac{a_{n+1}}{a_n} \le r < 1, \qquad \text{for all } n \ge N. \tag{56.1}$$

Then

$$a_{N+1} \le r a_N,$$

$$a_{N+2} \le r a_{N+1} \le r^2 a_N,$$

$$a_{N+3} \le r a_{N+2} \le r^3 a_N,$$

and so on. Thus we obtain a bound for the remainder (error) after summing N terms of the series:

$$\sum_{n=N+1}^{\infty} a_n = a_{N+1} + a_{N+2} + a_{N+3} + \cdots$$

$$\le a_N(r + r^2 + r^3 + \cdots)$$

$$= \frac{a_N r}{1 - r}, \tag{56.2}$$

by the formula for summing a geometric series. If a_N and r are both small, (56.2) should be much smaller, which is what we want. (Note that, unlike the methods discussed in Chaps. 52 and 54, no "correction" term is being added to the sum of the first N terms—yet.)

How can we determine a number r and a number N so that (56.1) is satisfied for all $n \ge N$? Well, the ratio test is going to be used only when the ratio of consecutive terms is a relatively simple function of n. It happens that the sequence of ratios is often a decreasing sequence, i.e.,

$$\frac{a_{n+1}}{a_n} \geq \frac{a_{n+2}}{a_{n+1}},$$ (56.3)

and furthermore that this is obvious once the ratio is written down. When that is the case, (56.1) will be satisfied with $r = a_{N+1}/a_N$, and then the error estimate (56.2) becomes

$$\sum_{n=N+1}^{\infty} a_n \leq \frac{a_{N+1}}{1 - a_{N+1}/a_N}.$$ (56.4)

The right-hand member of (56.4) can be computed for each N as the sum is being accumulated, and when it is less than the desired ϵ, the accumulation is stopped.

Example If $a_n = 1/n!$, then $a_{n+1}/a_n = 1/(n + 1)$, which is obviously a decreasing function of n. Since $\Sigma_0^{\infty} a_n = e$, inequality (56.4) says that

$$\left| e - \sum_{n=0}^{N} \frac{1}{n!} \right| \leq \frac{1}{N \cdot N!}.$$

Verify!

56.2 AN ALGORITHM

For most of the series for which the ratio test is appropriate, it is convenient to compute the nth term from the $(n - 1)$-th term according to some functional relationship, i.e., $a_n = f(n, a_{n-1})$. For example, if $a_n = 1/n!$, then $a_n = a_{n-1}/n$, so $f(x, y) = y/x$. By incorporating this kind of recursive definition of terms into our algorithm, we can write a program that works for many different series evaluations by merely changing a function definition statement. The first term of the series (which could be numbered 0, or 1, or something else) could be any number, perhaps the value of an input variable, but to keep things simple, let's suppose the first term is 1 and that it is numbered $n = 0$. (The first term could always be factored out and the series renumbered, if necessary.)

A slight modification of the discussion in Section 56.1 is now appropriate. Recall that in convergence problems "only the tail matters," so it is enough for (56.3) to apply for sufficiently large n. That is, it is not important for the ratios to decrease at the outset, as long as they decrease eventually. Furthermore, when two consecutive terms of the series are equal, their ratio is not less than 1, and the error analysis does not apply. [Also, the computer chokes on the right-hand side of (56.4).] Thus our algorithm requires a new wrinkle: a decision to bypass the end test unless $a_{N+1} < a_N$, so that the end test can be meaningfully computed.

These features of the algorithm, and the end test of Section 56.1, are incorporated in the flowchart shown in Figure 56.1, in which T represents a term, R the next term, and S the partial sum.

Exercises

1. Write a program to implement the flowchart. Output should include the sum and the number of terms required to get it. Take ϵ to be $5 \cdot 10^{-7}$.

Figure 56.1

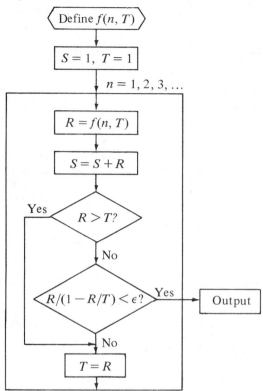

2. The preceding discussion was based on the assumption that the terms of the series are all positive. What modifications would you have to make in your program in order to allow the terms to be either positive or negative?

56.3 APPLICATION: e^x

In an earlier discussion of polynomial approximation (Sect. 45.3), we indicated how to compute e^{-x} for $x \geq 0$ by what you will now recognize as partial sums of the series $\Sigma(-1)^n x^n/n!$. That happens to be a rather bad way to proceed when there is a positive-term series for computing e^x, and $e^{-x} = 1/e^x$. This is because accumulation of many terms of an alternating series leads to loss-of-significance problems, and thereby limits the accuracy that can be achieved, even when the error estimate says that a reasonable number of terms suffice.

For any fixed value of x (in particular, for any positive x),

$$e^x = \sum_{n=0}^{\infty} \frac{x^n}{n!}. \qquad (56.5)$$

The ratio of consecutive terms in this series is

$$\frac{a_{n+1}}{a_n} = \frac{x^{n+1}n!}{x^n(n+1)!} = \frac{x}{n+1},$$

which is decreasing as a function of n. Thus inequality (56.4) tells us that

$$\left| e^x - \sum_{n=0}^{N} \frac{x^n}{n!} \right| \le \frac{x^{N+1}}{(N+1-x)N!}, \qquad N > x - 1. \qquad (56.6)$$

Notice that this error estimate is not any better than the "next-term" estimate one gets for the alternating series for e^{-x}; in fact, it is worse for large values of x. However, loss-of-significance problems are completely avoided by using the positive-term series. (For a more complete discussion of those problems, see Forsythe, Sect. 4.)

Exercise

3. Write a program to tabulate e^x and e^{-x} for x ranging from 0 to 10 in steps of 0.5. The values of e^x are to be computed to five significant digits (*not* five decimal places).

SUGGESTIONS: Your program from Exercise 1 can be adapted to this task. Enclose that program in a loop indexed for the appropriate values of x. Take $f(n, T) = xT/n$. (Why?) Instead of using a fixed ϵ, compute ϵ as $5 \cdot 10^{-6} \cdot S$. (Why?) Compute e^{-x} as $1/e^x$.

56.4 THE ROOT TEST [O]

Another convergence test, based on considerations similar to those previously mentioned, is the root test, which says that if $\lim_{n \to \infty} \sqrt[n]{|a_n|} < 1$, then $\Sigma \, a_n$ converges (absolutely). This is particularly useful if a_n involves nth powers. As in Section 56.1, we restrict our attention to positive-term series, and suppose that

$$\sqrt[n]{a_n} \le r < 1, \qquad (56.7)$$

for all $n > N$. Then $a_n \le r^n$ for the same values of n, and

$$\sum_{n=N+1}^{\infty} a_n \le \sum_{n=N+1}^{\infty} r^n = \frac{r^{N+1}}{1 - r}. \qquad (56.8)$$

Since the left-hand member of (56.8) is the remainder after summing N terms, the right-hand member is a bound for the error. In particular, if $\sqrt[n]{a_n}$ is decreasing for $n > N$, and $a_{N+1} < 1$, then $a_{N+1}^{1/(N+1)}$ may be taken as r, so the error estimate becomes

$$\sum_{n=N+1}^{\infty} a_n \le \frac{a_{N+1}}{1 - (a_{N+1})^{1/(N+1)}}. \qquad (56.9)$$

Exercise

4. Write a program to evaluate a series for which the root test is appropriate, to within ϵ. For this program a function definition describing the nth term explicitly is appropriate.

56.5 AN ACCELERATION ALGORITHM [O]

In Section 56.1 we used the ratio test to find a *bound* for the error after summing N terms of a series, as expressed in inequality (56.4). However, the right-hand member of (56.4) can also be taken as an *estimate* of the sum it bounds, i.e., of the error. Adding that estimate to the partial sum produces a "corrected partial sum," just as with the convergence acceleration methods discussed in Chapters 52 and 54. This can be expected to work well only if the inequality in (56.4) is close to being an equality. From the derivation it is clear that if the series is *exactly* geometric from term N on, then equality will hold. Thus if the series is "close" to being geometric (whatever that means), then the estimate of the error should be very good, and the corrected partial sum should be very close to the exact sum of the series. We will not attempt a careful or complete analysis of this technique here (see the references), but you may find it of interest to compare corrected and uncorrected partial sums in the next laboratory session. Do the former really "converge" more quickly, in the sense that digits do not change? Do they converge to the same numbers? And in the cases for which you have another way to compute the answer, do the corrected sums converge to the *right* answer?

To bring the discussion into line with standard acceleration methods, we make a slight change in the derivation of the error estimate. In Section 56.1 we estimated the ratio r for the "nearly geometric" tail by a_{N+1}/a_N. If this ratio is nearly constant, then we can estimate it just as well by a_{N+2}/a_{N+1}, but we choose to do this only in the denominator of (56.2)—i.e., we use two different estimates of r based on three consecutive terms of the series. This leads to a correction term of the form

$$\frac{a_{N+1}}{1 - a_{N+2}/a_{N+1}}. \tag{56.10}$$

If we simplify the compound fraction and change the sign of the denominator, (56.10) can be written as

$$-\frac{a_{N+1}^2}{a_{N+2} - a_{N+1}} \tag{56.11}$$

Correction of partial sums by terms of the form (56.11) is a classical acceleration technique called *Aitken's* Δ^2 (pronounced "delta square") *process*.

Alexander C. Aitken (1895–1967), who held the Chair in Mathematics at Edinburgh for some 30 years, introduced his acceleration process in connection with an extensive improvement on a method of Daniel Bernoulli for solving polynomial equations. The sequences of approximations in that application happened to be "nearly geometric" in a sense that could be made quite precise. However, the method has also turned out to be effective for a variety of other applications. For example, even though our derivation (limited to positive term series) obscures any reason for it, Δ^2 is remarkably effective at accelerating convergence of *alternating* series. It has also become the precursor of several families of more powerful "nonlinear" accelerators, all of which reduce to Δ^2 in their simplest cases. (See Shanks for some historical background in the context of introducing a major family of transforms, and see the two papers by Smith and Ford for extensive comparisons and tests of many of these acceleration methods, including the Shanks transforms.)

The "Δ^2" part of the name requires some explanation. As you know from your study of calculus, the Greek capital delta Δ stands for "difference." In the context of a sequence $\{S_n\}$, it stands for "difference of consecutive terms." Thus

$$\Delta S_n = S_{n+1} - S_n. \tag{56.12}$$

Now if the sequence in question is the sequence of partial sums of a series $\Sigma\, a_k$, then the difference between two consecutive partial sums is just a term of the series: $\Delta S_n = a_{n+1}$. The correction term (56.11) can be rewritten in this notation as

$$-\frac{(\Delta S_N)^2}{\Delta S_{N+1} - \Delta S_N}. \tag{56.13}$$

Finally, note that the denominator in (56.13) is also a difference, this time of terms of the sequence $\{\Delta S_n\}$ (which are just terms of the series). We abbreviate $\Delta(\Delta S_n)$ by $\Delta^2 S_n$. Thus the sequence of corrected partial sums, $\{T_n\}$, is computed from the sequence of partial sums by

$$T_N = S_N - \frac{(\Delta S_N)^2}{\Delta^2 S_N}. \tag{56.14}$$

It is this form for the correction term, with two different kinds of "squares of deltas," that gives rise to the name. However, there is no good reason to compute from this formula; using either (56.10) or (56.11) is more efficient. Can you think of a reason to prefer one of these to the other?

A final observation about differences: If $\{S_n\}$ is *any* sequence, then $a_{n+1} = \Delta S_n$ defines terms of a series $\Sigma\, a_k$ whose partial sums are the given S_n's. The sequence has a limit if and only if the series has a sum, so convergence of sequences and convergence of series are not two ideas, but are only one. The connecting link is the difference operator Δ, and the "inverse" operator (the connecting link in the other direction) is the summation operator Σ. Indeed, there is an entire discrete "calculus" of sums and differences that almost exactly parallels the continuous calculus of integrals and derivatives. Perhaps this will help you remember the distinction between the sequence of *terms* of a series and the sequence of *partial sums* of the same series. They are related in the "same way" as a derivative is related to a function or as a function is related to its antiderivatives.

Exercise

5. Modify the program from Exercise 1 to use (56.10) or (56.4) as a correction term instead of an end test. Stop the iteration when two consecutive corrected partial sums are within ϵ of each other. Put the output statement inside the loop, and tabulate terms, partial sums, and corrected partial sums.

References

Aitken; Burden, Faires, and Reynolds (pp. 50–53); Forsythe; Grabiner; Henriksen and Lees (Sect. 28); Hummel and Seebeck; Johnson and Riess (pp. 157–160); Kaplan and Lewis (Sects. 8–10); Shanks; Smith and Ford (1979, 1982).

CHAPTER 57

MORE SERIES EVALUATIONS [L]

Each of the following series is to be evaluated by using the appropriate program from Chapter 56. In the first ten cases, you need to compute the ratio $|a_n/a_{n-1}|$ in order to determine f. Also verify that the ratios are decreasing.

Exercises

1. $\displaystyle\sum_{n=0}^{\infty} \frac{1}{n!}$

2. $\displaystyle\sum_{n=0}^{\infty} \frac{2^n}{n!}$

3. $\displaystyle\sum_{n=0}^{\infty} \frac{1}{1 \cdot 3 \cdot 5 \cdots (2n + 1)}$

4. $\displaystyle\sum_{n=0}^{\infty} \frac{1}{(2n + 1)!}$

5. $\displaystyle\sum_{n=0}^{\infty} \frac{2^n}{1 \cdot 3 \cdot 5 \cdots (2n + 1)}$

6. $\displaystyle\sum_{n=0}^{\infty} \frac{n!}{(4n + 1)!}$

7. $1 + \displaystyle\sum_{n=2}^{\infty} \frac{n!}{3 \cdot 7 \cdots (4n - 1)}$

8. $\displaystyle\sum_{n=0}^{\infty} \frac{n + 1}{n!}$

NOTE: The next two exercises require the modifications indicated in Exercise 2 of Chapter 56.

9. $\displaystyle\sum_{n=0}^{\infty} \frac{(-1)^n}{n!}$

10. $\displaystyle\sum_{n=0}^{\infty} \frac{(-1)^n}{(2n + 1)n!}$

NOTE: For the next two exercises, the program from Exercise 4 of Chapter 56 is more appropriate. Rewrite the series in Exercise 12 so that the indexing starts at zero, and use the computer to get a value for the zero-th term. Change your program if necessary, to make the zero-th term an input variable.

11. [O] $1 + \displaystyle\sum_{n=1}^{\infty} \frac{1}{n^n}$

12. [O] $\displaystyle\sum_{n=2}^{\infty} \frac{1}{(\ln n)^n}$

13. Run your program from Exercise 3 of Chapter 56 to construct a table of values of e^x and e^{-x}.

14. Infinite series can be used for evaluation of certain integrals when the Fundamental Theorem cannot be applied directly. For example, to evaluate $\int_0^1 e^{-x^2} dx$, one first writes a power series for e^{-x^2}, which may be obtained from (56.5) by replacing x by $-x^2$. Since this series is absolutely convergent for all x, it may be integrated term by term. Show that this leads to the series in Exercise 10. (If you have a numerical integration program available, use it to check the answer from Exercise 10.)

15. [O] Run your program from Exercise 5 of Chapter 56, using the series in Exercises 1 and 2 as test problems.

DEBUGGING HINT: If you used (56.10) for your correction term, your program needs to know *three* consecutive terms at each step in the loop, the one just added to the partial sum and two beyond that. This means more variables to keep track of and update. It may also call for modification of the DEF statement defining terms, especially if your BASIC allows only one argument for functions. You may find it simpler to work with (56.4) first, even though that does not give quite the same result as Aitken's Δ^2.

16. [O] Once your acceleration program is debugged, exercise it on a variety of series from this chapter *and* from Chapter 55 (alternating series), especially those that required more than ten terms for satisfactory convergence.

17. [O] Try to answer the following questions regarding your acceleration program:
  a. For what types of series do you see the greatest acceleration in terms of the same accuracy being obtained from fewer terms of the series?

NOTE: Keep in mind that Δ^2 uses two terms beyond the one counted by the loop number. Note that comparisons with series evaluations from Chapter 55 are made against another accelerator (a special case of the Euler transform) that is specifically designed for alternating series. Evaluations by partial sums only would require far more terms (see the first paragraph of Chap. 54 again).

  b. Are there any cases in which the transformed sequence seems to converge to a wrong answer, or to a different answer from the one obtained before, or not at all? (Keep in mind that we have not proved anything about its convergence, let alone its rate of convergence.)
  c. What happens if you apply your accelerator program to the *p*-series exercises in Chapter 53?
  d. How much difference does it make whether you take your correction term from (56.4) or (56.10)?
  e. Does it matter whether you program the correction as in (56.10) or as in (56.11)?

18. [O] If you know about arrays and subroutines in your programming language, try the following: Write a program to generate the partial sums of a series and save the resulting sequence in an array. Then apply Aitken's Δ^2 to that sequence to construct another sequence, and save it in an array, two steps shorter than the previous one (why?). Now repeat: Apply Δ^2 to the transformed sequence to construct another transformed sequence, and so on. Since each sequence is two steps shorter than the previous one, the process stops with a sequence that has only one or two terms. Print each of the sequences as it is generated. (A printer is recommended for this. If you have a wide carriage printer and are sufficiently clever about formatting, you may find it more interesting to print the sequences side by side.) Now try your program on the series in Exercises 1 and 2 of Chapter 55. Recall that these series would require thousands of terms for evaluation by partial sums. You should see each of your transformed sequences converging faster than the previous one. This "iterated" Δ^2 is the way the method is usually applied, and indeed, Aitken did it this way himself.

CHAPTER 58

PROJECTS TO DO ON YOUR OWN

Each section of this chapter describes very briefly one or more projects that follow up on ideas discussed earlier in the text. These projects may be assigned (or you, the student, may attempt them on your own initiative) at any time after completion of the referenced chapters. References that may be helpful in pursuing these projects are included in each section.

58.1 CLASSICAL FORMULAS FOR π (CHAPS. 12–14)

Perhaps the earliest explicit formula for π is the following infinite-product formula obtained by the French mathematician Viète (see Chap. 19, Exercise 4) in the late sixteenth century:

$$\frac{2}{\pi} = \sqrt{\frac{1}{2}} \sqrt{\frac{1}{2} + \frac{1}{2}\sqrt{\frac{1}{2}}} \sqrt{\frac{1}{2} + \frac{1}{2}\sqrt{\frac{1}{2} + \frac{1}{2}\sqrt{\frac{1}{2}}}} \sqrt{\cdots} \qquad (58.1)$$

In another century, the following formulas (along with the infinite-series methods discussed earlier) were discovered by the English mathematician John Wallis (1616–1703) and the Irish peer Lord Brouncker (1620?–1684), respectively:

$$\frac{\pi}{4} = \frac{2 \cdot 4 \cdot 4 \cdot 6 \cdot 6 \cdot 8 \cdots}{3 \cdot 3 \cdot 5 \cdot 5 \cdot 7 \cdot 7 \cdots}, \qquad (58.2)$$

$$\frac{\pi}{4} = \cfrac{1}{1 + \cfrac{1^2}{2 + \cfrac{3^2}{2 + \cfrac{5^2}{2 + \cdots}}}} \qquad (58.3)$$

None of these formulas is an effective way to compute π, but you might find it of interest to compute a number of terms of the implied sequences to see how close you can get to π. (Write your programs with very long loops and interrupt them when you have seen enough.)

You may also find it of interest to explore where these formulas came from and how they can be proved.

References

Boyer (1968) (Chaps. 16, 18); Schreiner (Chap. 3); Thomas and Finney (Sects. 16–10); Turnbull; "Vieta."

58.2 THE METHOD OF FALSE POSITION (CHAPS. 15, 16, 18, 19)

The title of this section refers to a root-finding method that falls somewhere between interval-halving and the Newton–Raphson method. It is a method of interest historically and also because of its connection with the idea of linear interpolation.

Here is the basic idea: Suppose f is a continuous function on $[a, b]$, and $f(a)f(b) < 0$, so there is a root in the interval. An approximation to the root r is obtained by approximating f by the linear function (chord of f) whose graph passes through $(a, f(a))$ and $(b, f(b))$. The root c of the linear function is easily found, and the sign of $f(c)$ determines whether r lies in $[a, c]$ or $[c, b]$. In either case, we have a smaller interval than $[a, b]$ containing r, and the process can be repeated as many times as necessary.

Draw a picture to illustrate the preceding paragraph. Then derive the following expressions for c:

$$c = a - \frac{f(a)(b - a)}{f(b) - f(a)} = b - \frac{f(b)(a - b)}{f(a) - f(b)}. \qquad (58.4)$$

Either of the two descriptions of c will do for the iteration in a program. The symmetry of the two expressions indicates that it doesn't matter whether c is obtained by "correcting" a or b.

Write a program to find roots by false position. What is a reasonable stopping rule? Must the method converge? Compare results from this method with those obtained by interval-halving and Newton's method.

Suppose f has a continuous second derivative that does not change sign on $[a, b]$. Then chords and tangent lines lie on opposite sides of the graph, which suggests finding *two* approximations that bracket the root, one by false position, the other by Newton's method. (Compare the integration method of Chap. 26, which was based on a similar idea.) What are the advantages and disadvantages of such a combined method, as compared with (1) false position alone; (2) Newton's method alone? Write such a program, and compare results with the other methods.

References

Hildebrand (pp. 572–574); Johnson and Riess (pp. 145–148); Kuo (pp. 87–89); Leinbach (Chap. 5); McNeary (pp. 99–102); Ralston and Rabinowitz; Shampine and Allen (Chap. 3).

58.3 HORNER'S METHOD (CHAPS. 18, 19)

We noted in Chapter 18 that application of the Newton–Raphson method to polynomial equations produces what is known as "Horner's method," but which is, in reality, a very ancient method for finding roots. Its study is of interest for two reasons:

1. It permits evaluation of the polynomial *and* its derivative directly from a list of coefficients, so function definitions are not required.

2. When a root r is found, it automatically gives the coefficients for the quotient polynomial after dividing by $x - r$. The latter feature makes it possible to proceed to find additional roots by working with a polynomial of lower degree, and without fear of accidentally discovering the same root over again, unless it is in fact a multiple root. (There is a price

for this, however: Many successive extractions of roots by this method lead to accumulation of round-off errors.)

Here is a brief description of the method: Consider an arbitrary polynomial of degree ($\leq$)n:

$$p(x) = a_n x^n + a_{n-1} x^{n-1} + a_{n-2} x^{n-2} + \cdots + a_1 x + a_0.$$

In the computer, we arrange a list of its coefficients:

$$a_n\, a_{n-1}\, a_{n-2} \cdots a_2\, a_1\, a_0.$$

For any fixed number, t, say, we compute another list

$$b_{n-1}\, b_{n-2}\, b_{n-3} \cdots b_1\, b_0\, R,$$

where, for each index k,

$$b_k = a_{k+1} + b_{k+1} t, \tag{58.5}$$

and $R = a_0 + b_0 t$. [Think of R as b_{-1}, and (58.5) applies for it as well.] Now you should verify that if $x - t$ is divided into $p(x)$ (by long division, say) to get a quotient $Q(x)$ and a remainder R, then the b's are precisely the coefficients of Q, i.e.,

$$p(x) = (x - t)Q(x) + R, \tag{58.6}$$

where

$$Q(x) = b_{n-1} x^{n-1} + b_{n-2} x^{n-2} + \cdots + b_1 x + b_0.$$

Also, it follows from (58.6) that $p(t) = R$, so this is a way of evaluating p at any number (without computing any powers). So far, what we have described is a method you may have learned in high school under the title "synthetic division."

If we now write equation (58.6) as

$$Q(x) = \frac{p(x) - p(t)}{x - t},$$

and take a limit as $x \to t$, we see that $p'(t) = Q(t)$. Thus we can compute $p'(t)$ by another polynomial evaluation applied to Q; we can compute

$$c_{n-2}\, c_{n-3} \cdots c_1\, c_0\, S$$

by

$$c_k = b_{k+1} + c_{k+1} t, \tag{58.7}$$

and S (or c_{-1}) will be $p'(t)$. Now, given any starting guess x_0 for a root r of p, the Horner scheme tells us how to evaluate $p(x_0)$ and $p'(x_0)$, hence also the next approximation by Newton's method, according to formula (18.2). This can be repeated until R is sufficiently close to 0 to indicate that the corresponding x_j is (approximately) the root r. At that point, the b's are the coefficients of the quotient $p(x)/(x - r)$, and the process starts over, with the b's replacing the a's and a new guess x_0.

Use this scheme to write a program to find (all?) the real roots of a polynomial with real coefficients. It will help with this project if you learn about *arrays* or *vectors* in your programming language. However, you could use, say, ten different variable names for the a's, ten for the b's, and so on, which would allow for polynomials up to degree 9. (For lower degree, just set the high-order coefficients equal to 0.)

References

Boyer (1968); Burden, et al. (pp. 55–58); Hildebrand (pp. 588–595); McNeary (pp. 122–124); Stiefel (pp. 92–96).

58.4 THE METHOD OF SUCCESSIVE APPROXIMATIONS (CHAPS. 18, 19)

In Section 18.2 we saw that Newton's method for solving $f(x) = 0$ could be described in terms of a recursive sequence of numbers, $x_0, x_1, x_2, \ldots$ computed from

$$x_{k+1} = g(x_k), \tag{58.8}$$

for a certain function g, which sequence would hopefully converge to a number r such that

$$r = g(r), \tag{58.9}$$

and then r would also satisfy $f(r) = 0$. For some root-finding problems it is preferable (and conceptually simpler) to rewrite $f(x) = 0$ in the form $x = g(x)$ (which can often be done for many different choices of g), pick some starting guess x_0, and compute a sequence of successive "approximations" to the root r by formula (58.8). You might check that if g is continuous and the sequence $\{x_k\}$ defined by (58.8) converges at all, then the number r that is the limit *must* satisfy (58.9).

In order to have an example in mind, consider the cubic polynomial $f(x) = x^3 + 2x^2 + 2x + 1$. It is easy to see that $f(-1) = 0$, and if you graph f you will see that there are no other roots. Here are some of the many ways $f(x) = 0$ can be rewritten in the form $x = g(x)$:

$$x = g(x) = -\frac{1}{2}x^3 - x^2 - \frac{1}{2}; \tag{58.10}$$

$$x = g(x) = -\frac{1}{2}x^2 - 1 - \frac{1}{2x}; \tag{58.11}$$

$$x = g(x) = -(2x^2 + 2x + 1)^{1/3}; \tag{58.12}$$

$$x = g(x) = -2 - \frac{2}{x} - \frac{1}{x^2}. \tag{58.13}$$

How many others can you think of?

Looking back at Section 18.2, you find that the discussion of convergence presented there did not depend on knowing the form of g explicitly, but only on the fact that g was continuously differentiable on some interval containing both x_0 and r. Thus we conclude that, if $|g'(x)| \le M < 1$ on such an interval, then the sequence (58.8) necessarily converges to r, and if $|x_{k+1} - x_k| < \epsilon$, then $|x_{k+1} - r| < M\epsilon/(1 - M)$.

To illustrate that some choices of g are better than others, check that there is such an $M < 1$ for the functions defined in (58.10), (58.12), and (58.13) on some interval containing $r = -1$, but no such M exists for (58.11). Which of the three possible choices would lead to the quickest convergence (i.e., smallest M)? Some possible answers: For (58.10), $M = \frac{2}{3}$ works on $[-\frac{4}{3}, -\frac{2}{3}]$. For (58.13), $M = \frac{8}{27}$ works on $(-\infty, -1]$.

It is obviously very easy to program this method for root finding. Indeed, the program looks very similar to that for Newton's method, except that only one function definition is needed, that for g. However, it is essential with this method to use the techniques of calculus to first determine a suitable g and a "magic number" M so that the accuracy of the answer can be estimated. Explore this idea with some of the root-finding problems presented earlier or others of your own devising.

References

Burden, et al. (Sect. 2.2); Clark; Ford; Forsyth; Johnson and Riess (Sect. 4.3); Klamkin; Rosenblatt and Rosenblatt (pp. P-137–P-141); Schreiner (Chap. 5).

58.5 LEAST SQUARES (CHAPS. 21, 22)

There is an important minimization problem that is quite different from those considered earlier in this book, but that arises in many different applications. Furthermore, it is a problem that is easily "solved" with the methods of your calculus course, but the solution may not be very meaningful unless you have some assistance with the arithmetic it entails. (A calculator is often used, but the computer makes a good substitute.) If you take any laboratory science course, any quantitative social science course, or any statistics course, you are virtually certain to encounter this problem.

Let's suppose you are doing a laboratory experiment to determine a relationship between two variables X and Y. Suppose your apparatus is such that you can control the value of X and can measure corresponding values of Y. You choose several values $X_1, X_2, \ldots, X_n$, perform the experiment n times, and record your observations $Y_1, Y_2, \ldots, Y_n$. Now you may have reason to believe that the true relationship is one of simple proportionality:

$$Y = kX, \tag{58.14}$$

perhaps from theoretical considerations, perhaps from plotting your observations. Even if (58.14) is true, your observed points (X_i, Y_i) will not all lie on a straight line, of course, due to experimental errors. The problem: What is the "best" estimate of k that can be obtained from the observed data?

The most popular, but not the only, answer to this question is that the best k is the one that minimizes the expression

$$\sum_{i=1}^{n} (Y_i - kX_i)^2, \tag{58.15}$$

i.e., the sum of the squares of the differences between observed values Y_i and "predicted" values kX_i. This "least-squares" principle was discovered by Gauss while he was still a teenaged schoolboy in the 1790s. Can you think of any reason for calling it "best"?

The expression in (58.15) is a simple quadratic function of k, say, $F(k)$. (The X_i's and Y_i's are just numbers, of course.) By setting $F'(k)$ equal to 0 and solving for k, you can find

$$k = \frac{\sum_{i=1}^{n} X_i Y_i}{\sum_{i=1}^{n} X_i^2}. \tag{58.16}$$

Furthermore, you can check that $F''(k) > 0$ for all k, so the value given by (58.16) definitely minimizes F. If there are a lot of data points (X_i, Y_i), formula (58.16) represents a nontrivial computation, but one that is easy to program. (If you want n to be an input variable so that you can vary the number of data points, it is helpful to know about the use of *arrays* in your programming language.)

A much more likely relationship between variables than (58.14) is a general linear relationship

$$y = ax + b. \tag{58.17}$$

[Even if the relationship is known to be nonlinear, (58.17) may provide a good approximation over some suitably small interval.] Here there are two parameters, a and b, to be determined, but the least-squares principle is the same: We want to minimize

$$\sum_{i=1}^{n} [y_i - (ax_i + b)]^2 \tag{58.18}$$

by a suitable choice of a and b. If we fix b (as yet unknown), and set $Y_i = y_i - b$, $k = a$, $X_i = x_i$, and apply the previous result, we see that, in order to get a minimum, we must have

$$a = \frac{\sum_{i=1}^{n} x_i(y_i - b)}{\sum_{i=1}^{n} x_i^2}. \tag{58.19}$$

On the other hand, if we fix a (also as yet unknown), and set $Y_i = y_i - ax_i$, $k = b$, $X_i = 1$, and apply the previous result, we see that, to get a minimum, we must have

$$b = \frac{\sum_{i=1}^{n} (y_i - ax_i)}{n}. \tag{58.20}$$

Equations (58.19) and (58.20) are simultaneous equations in a and b, and it happens that they have a unique solution, which further happens to actually minimize (58.18). You can find this solution by first rewriting (58.19) and (58.20) as

$$(\Sigma x_i^2)a + (\Sigma x_i)b = \Sigma x_i y_i, \tag{58.21}$$

$$(\Sigma x_i)a + nb = \Sigma y_i, \tag{58.22}$$

and then solving the two simultaneous linear equations. Again, it is a simple task to write a program that takes the data points (x_i, y_i) as inputs and gives you the "best" values of a and b as outputs.

For sample data for a variety of applications, see Rosenblatt and Rosenblatt; or use *real* data from your next physics or chemistry lab or from some table in your sociology or economics text.

References

Rosenblatt and Rosenblatt (pp. P-81–P-92); Thomas and Finney (Sect. 13–10).

58.6 ANTIDERIVATIVES (CHAPS. 26, 27 or 29, 30)

Given a function f (continuous, say), how do you find a function F such that $F'(x) = f(x)$? In other words, how do you find antiderivatives or indefinite integrals? You either have already studied or will soon study a number of analytic techniques for solving this problem for specific functions *f*, but these methods leave you with many more functions *f* for which the question is unanswered. On the other hand, the Fundamental Theorem of Calculus provides a slick (if somewhat obscure) answer for *every* continuous function:

$$F(x) = \int_a^x f(t)\, dt, \tag{58.23}$$

for any fixed number *a* in the domain of *f*. Using a numerical integration formula, such as the Trapezoidal Rule, Midpoint Rule, or Simpson's Rule, you can evaluate $F(x)$ for any fixed *x*. Write a program to tabulate $F(x)$ over some interval $[a, b]$, given $f(x)$. Try it out with a variety of functions. Graph your output. Compare the output with indefinite integrals computed analytically, in those cases for which you know how to do this.

Reference

Thesing and Wood.

58.7 ON BEYOND SIMPSON (CHAPS. 29, 30)

There are many numerical integration formulas that are slightly more sophisticated than the Trapezoidal, Midpoint, and Simpson Rules, but not substantially more difficult to use. You might find it interesting to explore some of these and compare the results with those obtained earlier.

For example, there is the "Corrected Trapezoidal Rule,"

$$\int_a^b f(x)\, dx \doteq \frac{h}{2}\, [y_0 + 2y_1 + \cdots + y_n] + \frac{h^2}{12}\, [y_0' - y_n'], \tag{58.24}$$

where $h = (b - a)/n$, $y_i = f(a + ih)$, and $y_i' = f'(a + ih)$. Like Simpson's Rule, (58.24) is exact if *f* is a cubic polynomial, but it is generally more accurate for a given *n*, and *n* is not required to be even (see Hart, Hamming, and Munro).

Then there are formulas obtained by fitting quadratic, cubic, and so on, polynomials, but overlapping the fitted curves instead of stringing them end-to-end (see Peters and Maley). For example, for quadratic and cubic fits, we have

$$
\int_a^b f(x)\, dx \doteq h \left[\sum_{i=0}^n y_i - \frac{5}{8}\, (y_0 + y_n) + \frac{1}{6}\, (y_1 + y_{n-1}) \right.
$$
$$
\left. - \frac{1}{24}\, (y_2 + y_{n-2}) \right] \tag{58.25}
$$

and

$$
\int_a^b f(x)\, dx \doteq h \left[\sum_{i=0}^n y_i - \frac{23}{36}\, (y_0 + y_n) + \frac{5}{24}\, (y_1 + y_{n-1}) \right.
$$
$$
\left. - \frac{1}{12}\, (y_2 + y_{n-2}) + \frac{1}{72}\, (y_3 + y_{n-3}) \right]. \tag{58.26}
$$

Still other formulas may be obtained by fitting higher-degree polynomials, or by combining formulas with some weighting scheme, or in other, more abstruse ways. We mention by title only the formulas associated with the names Newton–Cotes (of which Trapezoidal and Simpson's are special cases), Hardy, Weddle, Legendre-Gauss, Euler-Maclaurin [of which (58.24) is a special case], and Romberg. (See Burden, et al.; Hildebrand; Johnson and Riess; Ralston and Rabinowitz; or Stiefel.)

References

Burden, et al. (Chap. 4); Hamming (Chap. 2); Hart; Hildebrand (Chaps. 3 and 8); Johnson and Riess (Chap. 6); Munro; Peters and Maley; Ralston and Rabinowitz; Stiefel (Sect. 6.2).

58.8 ON BEYOND QUARTICS (CHAPS. 32–36)

The procedure used in Chapters 32 and 34 to derive numerical differentiation formulas by fitting quadratic and quartic polynomials can obviously be carried out for any even degree $2k$; and the error can be shown to be on the order of h^{2k}, where h is the spacing between points (see the Exercise and Reference in Chap. 36). However, it is not necessary to carry out the polynomial fits explicitly to derive higher-order formulas. Cell shows how to use a weighted average of "sliding quartic" fits to derive the following formulas that are exact for sixth- and eighth-degree polynomials, respectively,

$$y' \doteq \frac{1}{60h} (-y_0 + 9y_1 - 45y_2 + 45y_4 - 9y_5 + y_6); \tag{58.27}$$

$$y' \doteq \frac{1}{840h} (3y_0 - 32y_1 + 168y_2 - 672y_3 + 672y_5 \\ - 168y_6 + 32y_7 - 3y_8). \tag{58.28}$$

Explore the use of these formulas, and compare the results with those obtained earlier.

Also, look into such matters as what to do when you cannot conveniently get function values on both sides of the point at which you want the derivative; what to do when you want values of higher derivatives; and how to choose an "optimal" value of h.

References

Burden, et al. (Sects. 4.1, 4.2); Cell; Ralston and Rabinowitz; Stiefel (Sect. 6.1).

58.9 RUNGE–KUTTA METHODS (CHAPS. 48, 49)

In Section 48.3 we derived a "predictor-corrector" method for solution of initial-value problems of the form

$$y' = f(t, y), \qquad y = y_0 \quad \text{when } t = t_0. \tag{58.29}$$

The integration step in that method was based on the Trapezoidal Rule. We now explore what can be done by using Simpson's Rule instead.

We use the notation $t_1 = t_0 + \Delta t$, $t_2 = t_1 + \Delta t$, ...; and $y_1, y_2, \ldots$ denote the computed approximations to $g(t_1)$, $g(t_2)$, and so on, g being the unknown solution function. Since

$$g(t_2) = y_0 + \int_{t_0}^{t_2} f(t, g(t)) \, dt,$$

if we approximate the integral by Simpson's Rule and the values of g by computed y's, we can write

$$y_2 \doteq y_0 + \frac{\Delta t}{3} [f(t_0, y_0) + 4f(t_1, y_1) + f(t_2, y_2)]. \tag{58.30}$$

Suppose for the moment that we already know y_1. Then our problem in using (58.30) is that y_2 appears on the right as well as on the left. If we could "predict" an approximation z_2 to y_2, then we could use (58.30) as a "corrector" and write

$$y_2 = y_0 + \frac{\Delta t}{3} [y_0' + 4y_1' + f(t_2, z_2)], \tag{58.31}$$

where we have abbreviated $y_0' = f(t_0, y_0)$, $y_1' = f(t_1, y_1)$. Rather than predict z_2 linearly, we can take advantage of the four (presumably) known items of information y_0, y_1, y_0', y_1', and fit a *cubic* polynomial that will predict a value for z_2 at t_2. You may check that this prediction turns out to be

$$z_2 = 5y_0 - 4y_1 + 2(\Delta t)y_0' + 4(\Delta t)y_1'. \tag{58.32}$$

In the absence of anything better to do, suppose we get our y_1 from the original predictor-corrector method:

$$z_1 = y_0 + (\Delta t)y_0', \tag{58.33}$$

$$y_1 = y_0 + \frac{\Delta t}{2} [y_0' + f(t_1, z_1)]. \tag{58.34}$$

Then we can organize the computation of y_2 from y_0, according to (58.31), in the following steps (check the details):

$$k_1 = 2(\Delta t)f(t_0, y_0), \qquad z_1 = y_0 + \frac{k_1}{2};$$

$$k_2 = 2(\Delta t)f(t_1, z_1), \qquad y_1 = y_0 + \frac{k_1}{4} + \frac{k_2}{4};$$

$$k_3 = 2(\Delta t)f(t_1, y_1), \qquad z_2 = y_0 - k_2 + 2k_3;$$

$$k_4 = 2(\Delta t)f(t_2, z_2), \qquad y_2 = y_0 + \frac{1}{6}(k_1 + 4k_3 + k_4).$$

Finally, we shift our point of view to ignore t_1 and y_1 entirely, replace the subscript 2 by 1, and replace $2(\Delta t)$ by Δt. Then the preceding formulas become

$$k_1 = (\Delta t)f(t_0, y_0),$$

$$k_2 = (\Delta t)f\left(t_0 + \frac{\Delta t}{2}, y_0 + \frac{k_1}{2}\right),$$

$$k_3 = (\Delta t)f\left(t_0 + \frac{\Delta t}{2}, y_0 + \frac{k_1}{4} + \frac{k_2}{4}\right), \tag{58.35}$$

$$k_4 = (\Delta t)f(t_1, y_0 - k_2 + 2k_3),$$

$$y_1 = y_0 + \frac{1}{6}(k_1 + 4k_3 + k_4).$$

Formulas (58.35) are called the (*first*) *Runge–Kutta method,* after two German applied mathematicians whose work was done primarily in the early part of this century. Of course, once you know how to get y_1 from y_0, you can increase all the subscripts to get y_2 from y_1, y_3 from y_2, and so on.

There are several other similar sets of formulas that are called Runge–Kutta methods, but one is enough to illustrate the idea. Programming formulas (58.35) is straightforward and not substantially more difficult than programming the linear predictor-corrector method. Compare the results you get from (58.35) with those obtained in Chapter 49.

References

Burden, et al. (Sect. 5.4); Ralston and Rabinowitz; Simmons (Chap. 2, App. A); Stiefel (Sect. 6.31).

58.10 EPIDEMICS (CHAPS. 50, 51)

The optional exercises in Chapter 51 may be considered as "projects" in the sense of this chapter. They involve tampering with the differential equations in the FOXRAB program so that the program simulates situations other than predator and prey populations. The main purpose in providing the FOXRAB program in "black-box" form was to produce the graphical output, which is possible only with somewhat more sophisticated programming than you have been expected to learn so far. You may want to look into such matters on your own, but it may be more instructive at this stage if you LIST a copy of FOXRAB and study the portions of it that do *not* have to do with graphical output. After such study, you should be able to write your own program that will at least tabulate solutions to more or less arbitrary systems of first-order differential equations.

In this section we look at differential equation models for the spread of epidemics. (Similar models are used in sociology for the spread of rumors.) Some of these models can be handled by FOXRAB by simply changing the differential equations. The more sophisticated models, however, have three or four differential equations, and therefore require a different program. Additional projects of this kind with applications in economics and political science can be found in Leinbach.

Suppose we have a large, homogeneous population of size n that tends to mix a lot in its daily contacts. (Think of an idealized central core of a large city.) Suppose further that at time $t = 0$, x_0 people are susceptible to a certain infectious disease and y_0 people actually have it. We require that y_0 should be small, and $x_0 + y_0 = n$. At any later time t, there will be some number x of susceptibles, some number y of infecteds, and some number z of others, called "removals," who have been quarantined or cured, or perhaps have died. We assume that $x + y + z = n$ for all t, and a possible model for the rates at which the three subpopulations change is

$$x' = -axy,$$
$$y' = axy - by, \qquad\qquad (58.36)$$
$$z' = by.$$

[Why are equations (58.36) reasonable? Exactly what assumptions are being made about the progress of the disease?] From a model like (58.36), a good deal of information can be obtained analytically (see Bailey or Maki–Thompson). For example, unless $x_0 > b/a$, the "relative removal rate," y' will never be positive, and no epidemic will develop. This leads to the obvious conclusion that epidemics can be prevented by inoculations (removing susceptibles before $t = 0$), even if the disease is present in the population. It also leads to the less obvious conclusion that a sudden influx of new people (increasing both n and x_0 "at" $t = 0$) can create an epidemic where none was possible previously. One can also learn from (58.36) that if $x_0 > b/a$, the total number of people who eventually get the disease is approximately $2 [x_0 - (b/a)]$.

Equations (58.36) can be solved numerically by FOXRAB by using only two of the three equations. The third is redundant, since $x' + y' + z' = 0$, an immediate consequence of the fact that $x + y + z$ is constant.

More sophisticated models would allow for some or all of the following possibilities: addition of new susceptibles due to births or immigration (which would be reflected by different kinds of terms in x'); additional removals due to deaths from other causes, or a continuing inoculation program (representing x-to-z transitions without going through y), or emigration (which could apply to any of the three subpopulations); additional susceptibles or infecteds from the removal population, if the disease is such that reinfection is possible. For some diseases, it would also be more realistic to include a fourth subpopulation consisting of those who have been infected but who are not yet infectious. [Equations (58.36) assume that the incubation time is zero.]

You have the power at your finger tips to explore any combination of the factors previously mentioned by writing down an appropriate system of differential equations and tabulating the solutions by the improved Euler method. Go to it!

References

Bailey (Chap. 9); Defares and Sneddon (Sect. 10.16); Leinbach (Chap. 7); Maki and Thompson (Chap. 9).

BIBLIOGRAPHY

The following abbreviations are used for the most frequently cited journals:

AMM	*American Mathematical Monthly*
MT	*Mathematics Teacher*

Abercrombie, T.J. The Sword and the Sermon. *National Geographic* 142 (1972), 3–45.

Aitken, A.C. "On Bernoulli's Numerical Solution of Algebraic Equations." *Proceedings of the Royal Society of Edinburgh* 46 (1926), 289–305.

Alfred, Brother U. "Exploring Fibonacci Numbers." *Fibonacci Quarterly* 1(1) (1963a), 57–63.

———. "Dying Rabbit Problem Revived." *Fibonacci Quarterly* 1(4) (1963b), 53–56.

Ayoub, R. "Euler and the Zeta Function." *AMM* 81 (1974), 1067–1086.

Bailey, N.T.J. *The Mathematical Approach to Biology and Medicine.* John Wiley & Sons, Inc., New York, 1967.

Ball, W.W.R. *Mathematical Recreations and Essays,* 11th ed. The Macmillan Company, New York, 1942.

Ballantine, J.P. "Elementary Development of Certain Infinite Series." *AMM* 44 (1937), 470–472.

Beckmann, P. *A History of π (Pi),* 2nd ed. Golem Press, Boulder, CO, 1971.

Bell, S., et al. *Modern University Calculus.* Holden-Day, San Francisco, 1966.

Bellman, R., and K.L. Cooke. *Modern Elementary Differential Equations,* 2nd ed. Addison-Wesley Publishing Company, Inc., Reading, MA, 1971.

Boas, R.P. "Estimating Remainders." *Mathematics Magazine* 51 (1978), 83–89.

Boyer, C.B. *A History of Mathematics.* John Wiley & Sons, Inc., New York, 1968.

———. "The History of the Calculus." In *Historical Topics for the Mathematical Classroom,* Thirty-first Yearbook of the National Council of Teachers of Mathematics. NCTM, Washington, DC, 1969. Reprinted in *The Two-Year College Mathematics Journal,* 1(1) (1970), 60–86.

Breusch, R. "A Proof of the Irrationality of π." *AMM* 61 (1954), 631–632.

Burden, R.L., J.D. Faires, and A.C. Reynolds. *Numerical Analysis,* 2nd ed. Prindle, Weber & Schmidt, Boston, 1981.

Bush, K.A. "On An Application of the Mean Value Theorem." *AMM* 62 (1955), 577–578.

Cairns, W.D. "Napier's Logarithms As He Developed Them." *AMM* 35 (1928), 64–67.

Cajori, F. "Historical Note on the Newton-Raphson Method of Approximation." *AMM* 18 (1911), 29–32.

———. "History of the Exponential and Logarithmic Concepts." *AMM* 20 (1913), 5–14, 35–47, 75–84, 107–117, 148–151, 173–182, 205–210.

———. "Who Was the First Inventor of the Calculus?" *AMM* 26 (1919), 15–20.

Cell, J.W. "An Accurate Method for Obtaining the Derivative Function from Observational Data." *AMM* 46 (1939), 87–92.

Clark, C. "Some Socially Relevant Applications of Elementary Calculus." *The Two-Year College Mathematics Journal* 4 (Spring 1973), 1–15.

Coate, G.T. "On the Convergence of Newton's Method of Approximation." *AMM* 44 (1937), 464–466.

Cohen, I.B. "Isaac Newton." *Scientific American* 193 (December 1955), 73–80. Reprinted in *Mathematics in the Modern World,* W.H. Freeman and Company, San Francisco, 1968.

Cohn, J.H.E. Letter to the Editor. *Fibonacci Quarterly* 2 (1964), 108.

Coolidge, J.L. "The Number *e*." *AMM* 57 (1950), 591–601. Reprinted in *Selected Papers on Calculus,* Mathematical Association of America, Washington, 1969, pp. 8–19.

———. "The Story of Tangents." *AMM* 58 (1951), 449–462.

———. "The Lengths of Curves." *AMM* 60 (1953), 89–93.

Coxeter, H.S.M. "The Golden Section, Phyllotaxis, and Wythoff's Game." *Scripta Mathematica* 19 (1953), 135–143.

Darst, R.B. "Simple Proofs of Two Estimates for *e*." *AMM* 80 (1973), 194.

Dean, R.A. "Vectors Point Toward Pisa." *The Two-Year College Mathematics Journal* 2(2) (1971), 28–39.

Defares, J.G., and I.N. Sneddon. *An Introduction to the Mathematics of Medicine and Biology,* North-Holland, Amsterdam, 1960.

Dorn, W.S., G.G. Bitter, and D.L. Hector. *Computer Applications for Calculus.* Prindle, Weber & Schmidt, Inc., Boston, 1972.

Eves, Howard. "Naperian Logarithms and Natural Logarithms." *MT* 53 (1960), 384–385.

———. "The Latest About π." *MT* 55 (1962), 129–130.

———. *An Introduction to the History of Mathematics,* 5th ed. Saunders College Publishing, Philadelphia, 1983.

Feldmann, R.W., Jr. "The Cardano–Tartaglia Dispute." *MT* 54 (1961), 160–163.

Ford, L.R. "Solutions of Equations by Successive Approximations." *AMM* 32 (1925), 272–285.

Forsyth, C.H. "The Solution of Certain Problems in Finance by the Method of Iteration." *AMM* 32 (1925), 126–129.

Forsythe, G.E. "Pitfalls in Computation, or Why a Math Book Isn't Enough." *AMM* 77 (1970), 931–956.

Franta, W.R. "Computer Arithmetic." *MT* 64 (1971), 409–414.

Gandz, S. "Origin of the Term 'Algebra.' " *AMM* 33 (1926), 437.

Gardner, M. "About Phi, an Irrational Number That Has Some Remarkable Geometrical Expressions." *Scientific American* 201 (August 1959), 128–134.

———. "Incidental Information About the Extraordinary Number π." *Scientific American* 203 (July 1960a), 154–162.

———. "Diversions That Involve the Mathematical Constant '*e*'." *Scientific American* 205 (October 1960), 160–168.

———. "Some Paradoxes and Puzzles Involving Infinite Series and the Concept of Limit." *Scientific American* 211 (November 1964), 126–133.

———. "The Multiple Fascinations of the Fibonacci Sequence." *Scientific American* 220 (March 1969), 116–120.

———. "The Abacus: Primitive but Effective Digital Computer." *Scientific American* 222 (January 1970), 124–127.

———. "Further Encounters with Touching Cubes, and the Paradoxes of Zeno as 'Supertasks.' " *Scientific American* 225 (December 1971), 96–99.

———. "The Calculating Rods of John Napier, the Eccentric Father of the Logarithm." *Scientific American* 228 (March 1973a), 110–113.

———. "On Expressing Integers as the Sum of Cubes and Other Unsolved Number-Theory Problems." *Scientific American* 229 (December 1973b), 118–121.

Gaughan, E.D. *Sequences and Limits.* Scott, Foresman and Company, Glenview, IL, 1972.

Giesy, D.P. "Still Another Proof of the Formula $\Sigma \, 1/k^2 = \pi^2/6$." *Mathematics Magazine* 45 (1972), 148–149.

Goldstine, H.H. *The Computer from Pascal to von Neumann.* Princeton University Press, Princeton, NJ, 1972.

Gould, S.H. "The Method of Archimedes." *AMM* 62 (1955), 473–476. Reprinted in *Selected Papers on Calculus,* Mathematical Association of America, Washington, DC, 1969.

Grabiner, J.V. "Is Mathematical Truth Time-Dependent?" *AMM* 81 (1974), 354–365.

Halsted, G.B. "Pi in Asia." *AMM* 15 (1908), 84.

Hammer, P.C. "The Midpoint Rule of Numerical Integration." *Mathematics Magazine* 31 (1958), 193–195. Reprinted in *Selected Papers in Calculus,* Mathematical Association of America, Washington, DC, 1969, pp. 328–330.

Hamming, R.W. *Calculus and the Computer Revolution.* Houghton Mifflin Company, Boston, 1968.

Hart, J.J. "A Correction for the Trapezoidal Rule." *AMM* 59 (1962), 33–36.

Hatcher, R.S. "Some Little-Known Recipes for π." *MT* 66 (1973), 470–474.

Heineman, E.R. "Geometric Interpretation of Polynomial Approximations of the Cosine Function." *AMM* 73 (1966), 648–649.

Henriksen, M. "Calculus and the Computer: A Conservative Approach." *Proceedings of a Conference on Computers in the Undergraduate Curricula,* University of Iowa, Iowa City, 1970, pp. 4.12–4.16.

Henriksen, M., and M. Lees. *Single Variable Calculus.* Worth Publishers, Inc., New York, 1970.

Higgins, G.A., Jr. "The Computer as a Tool for Building Mathematical Concepts and Historical Perspective." *Proceedings of a Conference on Computers in the Undergraduate Curricula,* University of Iowa, Iowa City, 1970, pp. 4.21–4.36.

Hildebrand, F.B. *Introduction to Numerical Analysis,* 2nd ed. McGraw-Hill Book Company, New York, 1974.

Hogben, L. *Mathematics for the Millions,* 3rd ed. W.W. Norton & Company, New York, 1953.

Hoggatt, V.E., Jr., and D.A. Lind. "The Dying Rabbit Problem." *Fibonacci Quarterly* 7 (1969), 482–487.

Horadam, A.F. "Fibonacci Sequences and a Geometrical Paradox." *Mathematics Magazine* 35 (1962), 1–11.

Hummel, P.M., and C.L. Seebeck. "A Generalization of Taylor's Expansion." *AMM* 56 (1949), 243–247.

Hurley, J.F. "An Application of Newton's Law of Cooling." *MT* 67 (1974), 141–142.

Johnson, L.W., and R.D. Riess. *Numerical Analysis,* 2nd ed. Addison-Wesley Publishing Company, Inc., Reading, MA, 1982.

Johnsonbaugh, R. "Summing an Alternating Series." *AMM* 86 (1979), 637–648.

Jones, P.S. " $\sqrt{2}$ in Babylonia and America." *MT* 42 (1949), 307–308.

———. "What's New About π?" and "Notes on Older Facts." *MT* 43 (1950), 120–122.

———. "Recent Discoveries in Babylonian Mathematics I: Zero, Pi, and Polygons." *MT* 50 (1957), 162–165.

Kammerer, H.M. "Sine and Cosine Approximation Curves." *AMM* 43 (1936), 293–294.

Kaplan, W., and D.J. Lewis. *Calculus and Linear Algebra,* Vol. I. John Wiley & Sons, Inc., New York, 1970.

Kaplan, W., R. Ritt, and R.R. Tebbs. "Applications of Elementary Differential Equations." Session III in *Proceedings of the Summer Conference for College Teachers on Applied Mathematics* (University of Missouri-Rolla, 1971), CUPM, Berkeley, 1973.

Kemeny, J.G. *Man and the Computer.* Charles Scribner's Sons, New York, 1972.

Kemeny, J.G., and T.E. Kurtz. *BASIC Programming,* 3rd ed. John Wiley & Sons, Inc., New York, 1980.

Kemeny, J.G., and J.L. Snell. *Mathematical Models in the Social Sciences.* Ginn-Blaisdell, Waltham, MA, 1962. Reprinted by The MIT Press, Cambridge, MA, 1972.

Klamkin, M.S. "Geometric Convergence." *AMM* 60 (1953), 256–258.

Klein, F. *Elementary Mathematics from an Advanced Standpoint*. The Macmillan Company, New York, 1932.

Kovach, L.D. "Ancient Algorithms Adapted to Modern Computers." *Mathematics Magazine* 37 (1964), 159–165.

Kuller, R.G. "Computer-Oriented Calculus." *Proceedings of a Conference on Computers in the Undergraduate Curricula,* University of Iowa, Iowa City, 1970, pp. 4.1–4.11.

Kuo, S.S. *Numerical Methods and Computers,* Addison-Wesley Publishing Company, Inc., Reading, MA, 1965.

Lanchester, F.W. "Mathematics in Warfare." In *The World of Mathematics,* ed. J.R. Newman, pp. 2136–2157. Simon & Schuster, Inc., New York, 1956.

Larrivee, J.A. "A History of Computers, I and II." *MT* 51 (1958), 469–473, 541–544.

Lehmer, D.H. "On Arccotangent Relations for π." *AMM* 45 (1938), 657–664.

Leinbach, L.C. *Calculus with the Computer*. Prentice-Hall, Inc., Englewood Cliffs, NJ, 1974.

Lotka, A.J. "Analytical Notes on Certain Rhythmic Relations in Organic Systems." *Proceedings of the National Academy of Sciences* 6 (1920), 410–415.

———. *Elements of Physical Biology*. The Williams & Wilkins Co., Baltimore, 1925. Reprinted as *Elements of Mathematical Biology*. Dover Publications, Inc., New York, 1956.

Maki, D.P., and M. Thompson. *Mathematical Models and Applications*. Prentice-Hall, Inc., Englewood Cliffs, NJ, 1973.

Malthus, T.R. "Mathematics of Population and Food." In *The World of Mathematics,* ed. J.R. Newman, Simon & Schuster, Inc., New York, 1956, pp. 1189–1199.

Matsuoka, Y. "An Elementary Proof of the Formula $\Sigma\, 1/k^2 = \pi^2/6$." *AMM* 68 (1961), 485–487.

McLaughlin, D.E. *Numerical Solution of Ordinary Differential Equations*. Iowa Curriculum Development Project, University of Iowa, Iowa City, 1973.

McNeary, S.S. *Introduction to Computational Methods for Students of Calculus,* Prentice-Hall, Inc., Englewood Cliffs, NJ, 1973.

Meserve, B.E. *Fundamental Concepts of Algebra*. Addison-Wesley Publishing Company, Inc., Reading, MA, 1953.

Miel, G. "Of Calculations Past and Present: The Archimedean Algorithm." *AMM* 90 (1983), 17–35.

Monzino, M.G. "On Simpson's Rule." *MT* 67 (1974), 165.

Morgan, L.A. "Areas by Infinite Series." *MT* 65 (1972), 754–756.

Morley, R.K. "The Remainder in Computing by Series." *AMM* 57 (1950), 550–551. Reprinted in *Selected Papers on Calculus*. Mathematical Association of America, Washington, DC, 1969, pp. 324–325.

Morrison, P., and E. Morrison. "The Strange Life of Charles Babbage." *Scientific American* 186 (April 1952), 66–73. Reprinted in *Mathematics in the Modern World,* W.H. Freeman and Company, San Francisco, 1968.

Mott, T.E. "Newton's Method and Multiple Roots." *AMM* 64 (1957), 635–638.

Munro, W.D. "Note on the Euler-Maclaurin Formula." *AMM* 65 (1958), 201–203.

Neugebauer, O. *The Exact Sciences in Antiquity*. Princeton University Press, Princeton, NJ, 1952. Second edition, Brown University Press, Providence, RI, 1957. Republished by Dover Publications, Inc., New York, 1969.

Newman, J.R. "Srinivasa Ramanujan." *Scientific American* 174 (June 1948). Reprinted in *Mathematics in the Modern World,* W.H. Freeman and Company, San Francisco, 1968.

Papadimitriou, I. "A Simple Proof of the Formula $\Sigma\, k^{-2} = \pi^2/6$." *AMM* 80 (1973), 424–425.

Parker, F.D. "Taylor's Theorem and Newton's Method." *AMM* 66 (1959), 51.

Pennisi, L.L. "Elementary Proof that e is Irrational." *AMM* 60 (1953), 474.

Peters, G.O., and C.E. Maley. "Numerical Integration over Any Number of Equal Intervals." *AMM* 75 (1968), 741–744.

Phillips, G.M. "Archimedes the Numerical Analyst." *AMM* 88 (1981), 165–169.

Pielou, E.C. *An Introduction to Mathematical Ecology.* Wiley-Interscience, New York, 1969.

Pinsky, M.A. "Averaging an Alternating Series." *Mathematics Magazine* 51 (1978), 235–237.

Pollack, S.V., and T.D. Sterling. *A Guide to PL/I.* Holt, Rinehart and Winston, Inc., New York, 1969.

Polya, G. *Mathematical Discovery,* vol. I. John Wiley & Sons, Inc., New York, 1962.

Porter, G.J. "An Alternative to the Integral Test for Infinite Series." *AMM* 79 (1972), 634–635.

Prielipp, R.W. "Niels Henrik Abel." *MT* 62 (1969), 482–484.

Ralston, A., and P. Rabinowitz. *A First Course in Numerical Analysis,* 2nd ed. McGraw-Hill Book Company, New York, 1978.

Read, C.B. "John Napier and His Logarithms." *MT* 53 (1960), 381–384.

———. "Shanks, Pi, and Coincidence." *MT* 60 (1967), 761–762.

Rescigno, A., and I.W. Richardson. "The Deterministic Theory of Population Dynamics." In *Foundations of Mathematical Biology,* ed. R. Rosen, vol. III. Academic Press, Inc., New York, 1973, Chap. 4.

Rosenblatt, L., and J. Rosenblatt. *Simplified BASIC Programming: With Companion Problems.* Addison-Wesley Publishing Company, Inc., Reading, MA, 1973.

Rosenthal, A. "The History of the Calculus." *AMM* 58 (1951), 75–86. Reprinted in *Selected Papers on Calculus,* Mathematical Association of America, Washington, DC, 1969.

Rosser, J.B. "Mathematics Courses in 1984." *AMM* 79 (1972), 635–648.

Sanford, V. "Brook Taylor (1685–1731)." *MT* 27 (1934a), 60–61.

———. "Colin Maclaurin (1698–1746)." *MT* 27 (1934b), 155–156.

Scarborough, J.B. "Formulas for the Error in Simpson's Rule." *AMM* 33 (1926), 76–82.

Schaumberger, N. "An Alternate Classroom Proof of the Familiar Limit for *e*." *The Two-Year College Mathematics Journal* 3 (Fall 1972), 72–73.

Schenkman, E. "The Independence of Some Exponential Values." *AMM* 81 (1974), 46–49.

Schepler, H.C. "The Chronology of π." *Mathematics Magazine* 23 (1950), 165–170, 216–228, 279–283.

Schoy, C. "Al-Bîrûnî's Computation of the Value of π." *AMM* 33 (1926), 323–325.

Schrader, D.V. "The Newton–Leibniz Controversy Concerning the Discovery of the Calculus." *MT* 55 (1962), 385–396.

Schreiner, A.T. *Computer Calculus.* Stipes Publishing Company, Champaign, IL, 1972.

Shampine, L.F., and R.C. Allen, Jr. *Numerical Computing: An Introduction.* W.B. Saunders Company, Philadelphia, 1973.

Shanks, D. "Non-Linear Transformations of Divergent and Slowly Convergent Sequences." *Journal of Mathematics and Physics* 34 (1955), 1–42.

Simmons, G.F. *Differential Equations.* McGraw-Hill Book Company, New York, 1972.

Smith, D.A. "Numerical Differentiation for Calculus Students." *AMM* 82 (1975), 284–287.

———. "Human Population Growth: Stability or Explosion?" *Mathematics Magazine* 50 (1977), 186–197.

———. "The Homicide Problem Revisited." *The Two-Year College Mathematics Journal* 9 (1978), 141–145.

———, and W.F. Ford. "Acceleration of Linear and Logarithmic Convergence." *SIAM Journal on Numerical Analysis* 16 (1979), 223–240.

———. "Numerical Comparisons of Nonlinear Convergence Accelerators." *Mathematics of Computation* 38 (1982), 481–499.

Smith, D.E. "Historical Survey of the Attempts at the Computation and Construction of π." *AMM* 2 (1895), 348–350.

Spencer, D.D. "Computers, Their Past, Present, and Future." *MT* 61 (1968), 65–75.

Stark, E.L. "Another Proof of the Formula $\Sigma\, 1/k^2 = \pi^2/6$." *AMM* 76 (1969), 552–553.

Stiefel, E.L. *An Introduction to Numerical Mathematics.* Academic Press, Inc., New York, 1963.

Stoller, D.S. *Operations Research: Process and Strategy*. University of California Press, Berkeley, 1964.

Stoneham, R.G. "A Study of 60,000 Digits of the Transcendental 'e'." *AMM* 72 (1965), 483–499.

Struik, D.J. "On Ancient Chinese Mathematics." *MT* 56 (1963), 424–432.

Te Selle, D.W. "Pi, Polygons, and a Computer." *MT* 63 (1970), 128–132.

Thesing, G.L., and C.A. Wood. "Using the Computer as a Discovery Tool in Calculus." *AMM* 81 (1974), 163–168.

Thomas, G.B., Jr. *Limits*. Addison-Wesley Publishing Company, Inc., Reading, MA, 1963.

———, and R.L. Finney. *Calculus and Analytic Geometry,* 5th ed., Addison-Wesley Publishing Company, Inc., Reading, MA, 1979.

Traub, J.F. "On Newton-Raphson Iteration." *AMM* 74 (1967), 996–998.

Turnbull, H.W. "The Great Mathematicians." In *The World of Mathematics.*, ed. J.R. Newman. Simon & Schuster, Inc., New York, 1956, Chaps. 5, 6.

Ulam, S.M. "Computers." *Scientific American* 211 (September 1964), 202–216. Reprinted in *Mathematics in the Modern World,* W.H. Freeman and Company, San Francisco, 1968.

Underwood, R.S. "Some Results Involving π." *AMM* 31 (1924), 392–394.

"Vieta." *MT* 23 (1930), 508 (illus., p. 466).

von Baravelle, H. "The Number e—the Base of Natural Logarithms." *MT* 38 (1945), 350–355.

———. "The Geometry of the Pentagon and the Golden Section." *MT* 41 (1948), 22–31.

———. "The Number π." *MT* 45 (1952), 340–348. Reprinted in same journal, 60 (1967), 479–487.

Weiner, L.M. "Note on Newton's Method." *Mathematics Magazine* 39 (1966), 143–145.

Wilf, H.S. "The Disk with the College Education." *AMM* 89 (1982), 4–8.

Wollan, G.N. "Maclaurin and Taylor and Their Series." *MT* 61 (1968), 310–312.

Wrench, J.W., Jr. "On the Derivation of Arctangent Equalities." *AMM* 45 (1938), 108–109.

———. "The Evolution of Extended Decimal Approximations to π." *MT* 53 (1960), 644–650.

CALCULUS TEXTBOOK CROSS-REFERENCE CHARTS

Each of the following charts shows the chapters of this book, grouped by topic, matched with appropriate sections of several standard calculus books. Some of the topics in this book are either not covered in the standard books or are covered too late to be included in a first-year course. In these cases, an appropriate point for coverage is indicated by a section or chapter number in parentheses. Where an advanced topic is taken up early in this book (e.g., sequences, polynomial approximations), the treatment is intended to be intuitive and to serve only as an introduction to more formal study later in the calculus course, not as a substitute. Of course, any topic in this book may be taken up at any convenient time later than the point indicated in the chart. In the case of polynomial approximations, each chart indicates two possible places for coverage: the section(s) covering the approximated functions (my preference) and the section (usually much later) covering Taylor's Theorem.

Chapter	Topic	Berkey	Anton	Edwards
0–5	Programming	(1.1–1.4)	(Chap. 1)	(1.1–1.3)
6	Functions, graphs	1.5	2.1	1.4
7–8	Sequences	(Chap. 2)	(Chap. 2)	(Chap. 1)
10	Derivatives	2.2, 3.1	3.1	1.6
12–14	Computation of pi	(Chap. 3)	(Chap. 2)	(Chap. 2)
15–17	Bisection, IVT	2.6	3.7	1.9
18–20	Newton's Method	4.6	4.9	3.4
21–23	Maxima/minima	4.3, 4.4	4.3	2.3
24–25	Upper, lower sums	6.2, 6.3	5.6, 5.7	4.2, 4.3
26–28	Trapezoidal Rule	6.7	9.9	4.7
29–31	Simpson's Rule	6.7	9.9	4.7
32–36	Differentiation	(Chap. 5 or 7)	(Chap. 6)	(Chap. 5)
37–38	Sine, cosine	6.6 or 12.4	5.9 or 11.10	4.5 or 11.2
39–40	Arctangent	10.5 or 12.4	8.3 or 11.10	7.3 or 11.2
41–43	Logarithms	9.2 or 12.4	7.2 or 11.10	6.2 or 11.2
45–47	Compound interest	9.5	7.7	6.5
48–49	Differential equations	5.10, 9.6	7.7	6.5–6.8
50–51	Predator-prey	(Chap. 10 or 11)	(Chap. 8)	(6.8)
52–53	Integral test	13.5	11.4	12.4
54–55	Alternating series	13.7	11.7	12.6
56–57	Ratio test	13.6	11.5	12.6

Chapter	Topic	Ellis	Gillett	Grossman
0–5	Programming	(1.1, 1.2)	(1.1–1.4)	(1.1–1.6)
6	Functions, graphs	1.3, 1.4	1.5	1.7
7–8	Sequences	(2.1)	(Chap. 2)	(2.1–2.4)
10	Derivatives	3.2	2.2	2.7–2.9
12–14	Computation of pi	2.2	(Chap. 3)	(2.2)
15–17	Bisection, IVT	2.5	4.2	2.10
18–20	Newton's Method	3.9	5.4	4.8
21–23	Maxima/minima	4.1	5.2, 5.5	4.5–4.7
24–25	Upper, lower sums	5.1	6.1, 6.2	5.3
26–28	Trapezoidal Rule	7.6	10.6	8.10
29–31	Simpson's Rule	7.6	10.6	8.10
32–36	Differentiation	(Chap. 8)	(Chap. 7)	(Chap. 3 or 8)
37–38	Sine, cosine	(5.5) or 9.1	6.5 or 11.5	7.2 or 13.3
39–40	Arctangent	6.6 or 9.1	9.1 or 11.5	7.3 or 13.3
41–43	Logarithms	5.7 or 9.1	8.1 or 11.5	6.3 or 13.3
45–47	Compound interest	9.2	8.6	6.6
48–49	Differential equations	6.5	7.7, 8.6	6.5
50–51	Predator-prey	(Chap. 7)	(Chap. 9)	(6.5)
52–53	Integral test	9.5	12.3	14.5
54–55	Alternating series	9.7	12.5	14.7
56–57	Ratio test	9.6	12.6	14.6

Chapter	Topic	Larson	Leithold	Mizrahi
0–5	Programming	(1.1–1.5)	(1.1–1.4)	(1.1–1.3)
6	Functions, graphs	1.6	1.5	1.4
7–8	Sequences	(Chap. 2)	(2.1)	(Chap. 2)
10	Derivatives	3.1, 3.2	3.2	3.2
12–14	Computation of pi	(Chap. 4)	(Chap. 2)	(Chap. 3)
15–17	Bisection, IVT	2.2	2.8	2.4
18–20	Newton's Method	5.5	(4.1)	11.5
21–23	Maxima/minima	5.1	4.3	4.3, 4.8
24–25	Upper, lower sums	6.3	6.1	5.3
26–28	Trapezoidal Rule	10.10	6.6	9.10
29–31	Simpson's Rule	10.10	6.6	9.10
32–36	Differentiation	(Chap. 7)	(Chap. 5)	(Chap. 6)
37–38	Sine, cosine	9.3 or 18.8	(Chap. 7) or 14.5	5.5 or 11.4
39–40	Arctangent	9.5 or 18.8	9.6 or 14.5	8.4 or 11.4
41–43	Logarithms	8.2 or 18.8	8.4 or 14.5	7.2 or 11.4
45–47	Compound interest	8.1	8.7	7.6
48–49	Differential equations	8.3	8.7	7.7
50–51	Predator-prey	(Chap. 10)	(8.7)	(Chap. 8)
52–53	Integral test	18.3	15.5	12.3
54–55	Alternating series	18.5	15.6	12.4
56–57	Ratio test	18.5	15.6	12.3

Chapter	Topic	Salas	Shenk	Stein
0–5	Programming	(1.1–1.5)	(Chap. 1)	(Chaps. 0, 1)
6	Functions, graphs	1.6	2.1, 2.2	1.3
7–8	Sequences	(2.1, 2.2)	(2.5, 2.6)	(Chap. 2)
10	Derivatives	3.1	3.1	3.2
12–14	Computation of pi	(Chap. 2)	(Chap. 2)	(Chap. 2)
15–17	Bisection, IVT	2.6	2.7, 2.8	2.7
18–20	Newton's Method	(3.9)	4.9	Chap. 4 or 10.13
21–23	Maxima/minima	4.3–4.5	4.6, 4.7	4.5
24–25	Upper, lower sums	5.1, 5.2	5.1, 5.2	5.3
26–28	Trapezoidal Rule	8.8	8.8	8.9
29–31	Simpson's Rule	8.8	8.8	8.9
32–36	Differentiation	(Chap. 4)	(Chap. 3 or 6)	(Chap. 6)
37–38	Sine, cosine	7.2 or 13.6	5.5 or 10.1	5.5 or 10.8
39–40	Arctangent	7.4 or 13.6	7.8 or 10.1	6.5 or 10.8
41–43	Logarithms	6.3 or 13.6	7.2 or 10.1	6.3 or 10.8
45–47	Compound interest	6.5, 6.6	7.6	6.1
48–49	Differential equations	6.6, 6.8	7.5, Chap. 9	6.7
50–51	Predator-prey	(6.8)	9.2	(Chap. 7)
52–53	Integral test	13.3	11.5	10.3
54–55	Alternating series	13.5	11.4	10.5
56–57	Ratio test	13.4	11.5	10.4

Chapter	Topic	Swokowski (A)	Swokowski (3)	Thomas
0–5	Programming	(1.1–1.3)	(1.1–1.3)	(1.1–1.4)
6	Functions, graphs	1.4	1.4	1.5
7–8	Sequences	(Chap. 2)	(Chap. 2)	(1.6–1.7)
10	Derivatives	3.2	3.2	1.8–1.9
12–14	Computation of pi	(Chap. 2)	(Chap. 2)	(Chap. 2)
15–17	Bisection, IVT	2.6	2.5	1.12
18–20	Newton's Method	3.10	3.9	2.10
21–23	Maxima/minima	4.6	4.6	3.5, 3.6
24–25	Upper, lower sums	5.2	5.2	4.6–4.8
26–28	Trapezoidal Rule	5.7	5.7	4.12
29–31	Simpson's Rule	5.7	5.7	4.12
32–36	Differentiation	(Chap. 6)	(Chap. 6)	(Chap. 5)
37–38	Sine, cosine	5.6 or 10.5	8.3 or 10.5	4.10, 6.1, or 12.2
39–40	Arctangent	8.4 or 10.5	8.5 or 10.5	6.3 or 12.3
41–43	Logarithms	7.2 or 10.5	7.4 or 10.5	6.4–6.6 or 12.3
45–47	Compound interest	7.5	7.6	6.12
48–49	Differential equations	7.6	7.7	6.12
50–51	Predator-prey	(Chap. 8)	(7.7)	(Chap. 7)
52–53	Integral test	11.3	11.3	11.5
54–55	Alternating series	11.4	11.4	11.7
56–57	Ratio test	11.5	11.5	11.5

INDEX

Page numbers in *italics* indicate illustrations; those followed by *t* indicate tables; and those followed by *n* indicate footnotes.